T0006114

ELECTRICAL AND MECHANICAL ENGINEERING 101

ELECTRICAL AND MECHANICAL ENGINEERING 101

AN ESSENTIAL GUIDE TO MASTERING THE SUBJECT

DAVID BAKER

Picture credits

All images David Baker with the exception of the following:

9 (top) Alastair Key and Melin Eran/University of Kent, (bottom) Ethan Doyle White, 11 Markus Hagenlocher, 14 Carlito20, 15 Knight's American Mechanical Dictionary, 25 Diebold Schilling/Adrian Michael, 27 Rama, 28 Kerina Yin, 29 Becarlson, 30 Cjp24, 46 Frankie Roberto, 52 US Library of Congress, 53 N A Parish, 58 Lead Holder, 59 Hiro36, 61 Keithonearth, 64 Aconcagua, 70 Simon Boddy, 74 Alleghany Ludlum Steel Corporation, 77 NASA, 78 Uddeholms AB, 80 Polyacrylonitrile, 82 Plasan, 83 NASA, 84 NASA, 86 Eric Garber, 88 RaBoe, 90R Johannes Maximilian, 90L Daniel Holth, 91 Alfred T. Palmer, 93 Ian Dunster, 95 Jeff Dahl, 96 Nubiter, 97 Nimbus 227, 98 K. Aainqatsi, 104 Steag, 105 Mgiardina099, 111 Sarah Adrita, 114 Dennis Jenkins, 120 NASA, 123 North American Aviation, 125 Ferdna, 129 NASA, 131 J. Brew, 132 Rama, 132 NASA, 138 LANL, 142 Jason Minshull, 143 US Department of Defense, 145 USAF, 150 Benjamin West, 153 Andrew Balet, 159 Paul Cuffe, 160 Clark Mills, 165 Siemens Pressebild, 166 US Dept of Energy, 168 Prokudin-Gorsky, 171 Sandia National Laboratories, 173 Christopher Ziemnowicz, 175 Oleg Alexandrov, 179 Jafet, via David Baker, 180 R. Dervisoglu, 181 North American Aviation, 185 Ambanmba, 186 Alessandro Nassiri, 187 Messer Woland, 188 Ralf Roletschek, 191, 194 US Air Force, 196 Jesper Olsson, 198 Massimiliano Lincetto, 201T Rolls-Royce, 201B NASA, 204 Michel Gilliand, 205, 207 NASA, 208 DoE, 210 NRL, 211 JET, 214 NASA, 215 KUKA AG, 216 Ocado.

This edition published in 2024 by Sirius Publishing, a division of Arcturus Publishing Limited,
26/27 Bickels Yard, 151–153 Bermondsey Street,
London SE1 3HA

ISBN: 978-1-3988-3688-4
AD011589US

Printed in China

CONTENTS

INTRODUCTION

Preceded by the era of toolmaking and the introduction of farming, engineering has been the third age of humankind in its evolution from the Stone Age to the world of machines, industrialization, telecommunications, and transportation. Engineers have made the world we live in today and created the technology that informs the lives of people around the globe. It is the purpose of this book to show how that has come about, to examine the various disciplines within the field of electrical and mechanical engineering, and to introduce the reader to the basic principles that underpin an activity that links physics, chemistry, mathematics, and technology.

An important part of understanding this world of complex and sophisticated machines is the way the entrepreneurs of the past forged natural materials into compounds and alloys, reforming them and exchanging levels of energy and producing heat engines. From this came an industrial revolution, founded in Britain and exported to the developed and developing world, making possible previously unimagined opportunities for all. In the following pages, several separate steps in the development of electrical and mechanical engineering are discussed with their respective origins and development paths. But this is not a history book.

The field of electrical engineering is growing fast, changing according to the evolving texture of human society. The extraordinary growth in human population levels is borne out by the unanticipated demands made on all these aforementioned capabilities, supporting a world in which it has taken all of human history to reach a population of 2 billion and the span of a human lifetime to quadruple that number. This has placed stress on the existing infrastructure and forced new and innovative technologies to come out of research and development in all fields of engineering.

With ever-demanding environmental concerns, the challenges now are immense. Issues that can translate into outstanding opportunities for new and aspiring engineers can be seen as daunting, or they can be grasped as routes to professional career fulfilment and a stimulus to ingenuity and innovation. Opportunities today are greater than they have ever been. There are new and highly specialized career development paths involving almost every facet of life, and the demand for improved and more efficient machines has never been greater. The retail and manufacturing worlds demand increasingly more sophisticated and effective solutions, and it is the task of the modern engineer to meet those demands in full.

The pioneering days are not over and in some respects are only just beginning. What in a previous generation took several decades to emerge can now be realized in weeks or months, and what corporations and big companies controlled in the past is now delivered by new start-up companies and truly creative thinkers. Be it in transforming the wasteful ways of the past into efficient and sustainable solutions for the future or in satisfying the needs of an environmentally conscious society, engineers will continue to contribute as enablers in a future world, building a better place for humans to occupy.

Chapter One

THE DAWN OF IDEAS—THE EMPIRICAL AGE

Early Engineering—Machines and Metalworking—Islamic Innovation—Chinese Inventions—Working Machines—The Role of Materials—The Forging of Global Connectivity

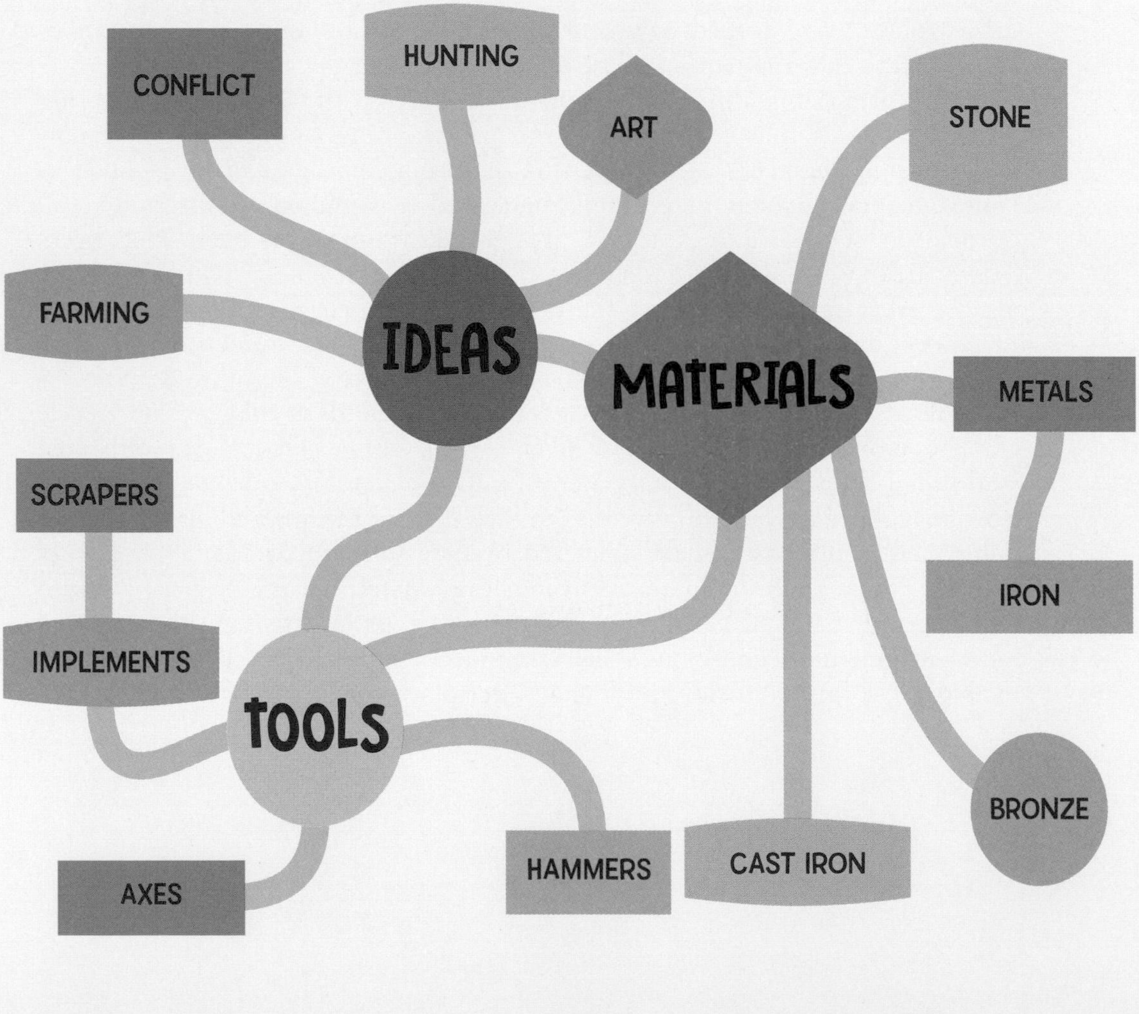

CONFLICT
HUNTING
ART
STONE
FARMING
IDEAS
MATERIALS
METALS
SCRAPERS
IRON
IMPLEMENTS
TOOLS
BRONZE
AXES
HAMMERS
CAST IRON

Stone tools from Skorba in Malta show the diversity of creative ingenuity.

EARLY ENGINEERING

Invention and discovery have gone hand in hand from the dawn of abstract thinking—activities buried deep in the ancient past. From the Paleolithic stone toolmakers of more than 3 million years ago to the application of known principles of engineering and science in the 21st century, the observation and use of natural laws and mechanical practicalities have underpinned all aspects of modern society. The collective application of laws, principles, and ideas have forged new ways of working through a process of test and evaluation. In fact, for several millennia that was the only way of fashioning natural materials into tools and working machines.

A stone axe head and Clovis spear point, perhaps the earliest examples of refashioned natural materials.

Long before the scientific application of calculation and extrapolation, hand tools were shaped to improve the ability of primitive humans to survive the rigors of a hostile environment and outwit predators. In fact, defense against a wide range of aggressive animals must have played a significant part in the development of those parts of the brain responsible for nurturing creative thinking. In turn, that would lead to a level of cognitive development where observation, experimentation, and evaluation served as a prerequisite for stored learning. And when that was achieved, humanity put its running shoes on!

However, the development of ***engineered structures***—and engineering per se—relied on need: the necessity to improve all aspects of the human condition, including the need for food, water, warmth, protection, and a means of defense against predators and other humans. In no specific order or sequence, six essential tools, or engineered items, were developed.

The precise order of application of these in terms of time is unknown, although some were given functional attributes as machines more than 2,500 years ago.

1. The wedge or ramp
2. The wheel and the central axle
3. The screw
4. The lever
5. The pulley
6. The crane

Neolithic farmers designed for a new and more settled existence.

The rudimentary tools and goods found in roundhouses fabricated from wattle and daub—techniques developed through abstract thinking.

The Archimedean screw as applied to a combine harvester in an integrated machine built on empirical principles.

Archimedes and his legacy

The best known architect of one of the earliest machines was Archimedes (288–212 BC), whose famous screw was developed in the 3rd century BC. This was applied to a wide range of tasks throughout the world and, in some countries, remains to this day the primary means of lifting water from one level to another. Archimedes bequeathed a device that was directly responsible for the replacement of sail with steam, propelling ships through one fluid—water—and propelling aeroplanes through another—air. Thus did the Archimedean screw connect the 3rd century BC to the 21st century AD and in so doing aid in the development of machines, albeit ones with limits defined by the ***balance of forces*** and not the ***application for work***, defined as the trade between force and distance.

What Archimedes also did was to explore the opportunities in mechanical advantage offered by the lever—work that inspired later Greeks to further define the various applications with a lever, windlass, pulley, wedge, and screw. But all of these applications avoided the dynamics of working machines and merely defined and exploited the static balance of force in simple machines.

What is a machine?

The technical definition of a simple machine is one in which the amount of power coming out (F_{out}) is equal to the force applied going in (F_{in}) and can be no greater. The mechanical advantage is frequently obtained in such simple machines by multiplying the magnitude of force by a specific factor: $MA = F_{out}/F_{in}$.

Clearly, a simple machine does not itself contain a source of energy other than that applied to input a force, and this defines the amount of work that it can do; without ***friction*** (or elasticity), this is known as an ***ideal machine***. In these, due to the conservation of energy, the power output (P_{out}), as defined through a rate of energy output, is equal to the power input (P_{in}). The power output equals the velocity of the load (v_{out}) multiplied by the load force ($P_{out} = F_{out}\, v_{out}$-) just as the input from the power side of the applied force is also equal to the velocity of the input point (v_{in}) multiplied by the applied force: $P_{in} = F_{out}\, v_{in}$-. Therefore, $F_{out} v_{out} = F_{in}\, v_{in}$.

Machines and monuments

While we benefit from having Greek text to explain motivations and draw conclusions for defining essential principles of force and work, previous civilizations a thousand years earlier exercised their own applications; people of the ancient Near East and city-states such as Mesopotamia and Babylon applied levers and balances. Indeed, applications date back 5,000 years as the first examples of machines were made to work in ways that had previously been impossible. In Britain, 4,500 years ago, these machines constituted levers and probably pulleys—perhaps even the yet-to-be-invented Archimedean screw. The complex geometry of megalithic monuments in the British Isles of the Neolithic and the Bronze Age testify to the universal application of these simple machines during the dawn of the ***empirical age***.

Tools developed with surprising sophistication were employed for complex shaping tasks, gouging mortise cavities for tenon joints fashioned in quartz sandstone, or the silcrete (sarsen) blocks now seen on a grand scale at Stonehenge. Other stone tools were used to fashion tongue and groove joints to stabilize lintel blocks that were slotted together. The application of stone tools to shape and fashion blocks weighing up to 27.5 tons (25 tonnes) invokes the use of mauls, hammer-stones, levers, and ramps—all applications of the simple machine—and may have involved cranes and pulleys.

We cannot know for sure the level of machine application to the construction

Shaping and fashioning for esthetic and purposeful applications, be it calendar or astronomical alignment, is a fine example of early empirical thinking.

of giant structures but we do know that, 4,500 years ago, the Egyptians used levers, pulleys, wheels on axles, cranes, rollers, and ramps, together with wedges, tooled shaping of natural materials, and perhaps other machines to build the pyramids. And in fact, beginning around the time Stonehenge was being erected in the grand finale of its multi-faceted form, almost all the simple machines were either in use or were being introduced tentatively in dispersed civilizations across the known world.

MACHINES AND METALWORKING

Simultaneous with the development of advanced tools and simple machines—which coalesced around farming and the need for irrigation, architectural constructions, and military engines for war-fighting purposes—ancient city-states in the Middle East began working in metals and other exotic materials.

The application of an Archimedean screw at Huseby Bruk, Sweden, is testament to an age-old principle of mechanical engineering.

This was paralleled by the emergence of the Bronze Age in central and northwestern Europe, from which would flow the Iron Age and the eventual manufacture of materials never found in nature. Such materials played an equally vital role in the development of ***advanced machines*** and ***shaped tools*** fabricated from the new materials.

Indeed, the use of metals was significant in the fabrication of tools and simple machines. Noted for its dominance in Mesopotamia for more than 2,500 years, before its demise early in the 7th century, the Assyrian Empire is also known for its fine metalwork and use of iron weapons. The Assyrian people were the first to build ***war machines*** of a type that would endure into the European medieval period, most notably battering rams, siege engines, and large catapults. These developments were paralleled by the Roman Republic and the later Roman Empire, which introduced a significant ***metalworking industry*** that accompanied their forces as they invaded foreign lands, and remain as a legacy in the cities that came in their wake.

Elsewhere, in the 1st century AD, Hero of Alexandria (AD 10–70) made the first ***steam-powered machine***—the aeolipile or "wind ball." This consisted of

Hero's aeolipile demonstrated action and reaction before they were ever defined.

a spherical water container, mounted in such a way that it was free to spin and fitted with release spouts on diametrically opposite sides of the sphere. The aeolipile was placed over a fire so the water inside would heat up, producing steam that was expelled from the spouts and provided positive ***thrust***, causing the container to rotate. Shortly afterwards, in China, the noted astronomer and scientist Zhang Heng (AD 78–139) perfected an improved water clock and went on to design a seismometer, in addition to a set of differential gears for chariots. These increasingly revolutionary devices, all simple machines, provided the impetus for the development of more complex and sophisticated devices over succeeding centuries. For example, of more esthetic satisfaction than any practical use, the first organ was assembled in the 3rd century BC but was greatly advanced as a sophisticated musical instrument in the 8th century. These simple machines—the foundation of engineering—underlie the structure for humankind's most creative skills and talents.

Had the collapse of the Roman Empire not occurred in the 5th century, it is likely that the advent of modern science and engineering would have produced a greater proliferation of working devices, and there is circumstantial evidence to suggest that some innovative, albeit simple, machines would have been added to the inventory. As it was, as Europe entered the Middle Ages (5th–15th centuries), Western machine technology stalled for more than 1,000 years, while further and more dramatic advances in the use of simple machines occurred in the Middle East, India, and China.

ISLAMIC INNOVATION

In what is now regarded as the Islamic Golden Age—roughly the end of the 6th century through the end of the 15th century—in the Middle East and throughout the Mediterranean world, great strides were made with simple machines and their evolution into what could be considered as mechanics, with applications in a new age of technology. This was in stark contrast to Western Europe, where there was little interest in maintaining the institutional infrastructure that had characterized the classical world of Greece and Rome. It can be postulated that had it not been for Islamic seats of learning and the entrenched belief among Muslims that knowledge was a precious gift to be nurtured and extended, much learning would have been lost. But it was not lost, and there was a gradual increase in mechanical use of simple machines, setting down the fundamental elements of basic mechanical principles.

The list of inventions that came out of this Golden Age is impressive and constitutes an important link with the European renaissance in mechanical engineering that underpinned the development of scientific thinking and supplanted empirical design. They included the invention by the Persians in the 5th–9th centuries of the windmill and the wind pump, the latter being a derivative of the windmill used for raising water by way of a mechanical application of wind power. From India came the cotton gin in the 6th century, a mechanical device for separating fibers from cotton seeds that involved a relatively complex system of wheels, gears, and levers. This required a significant degree of ***manufacturing*** of small and highly accurate components together with precision assembly, reliability, and robust design—but it worked!

The spinning wheel came into widespread use in the Middle East during the 11th century and, together with the cotton gin, created one of the first international industries, expanding on the trade and exchanges that had characterized the early expansion of toolmaking during the latter decades of the hunter-gatherer communities in Western Europe, and the introduction of farming and agriculture, growing crops and corralling livestock. This cross-fertilization, development, and co-ordination of ideas and inventions would later be the catalyst for inspiring Western scientists and engineers in the Age of Enlightenment (17th–19th centuries).

In fact, quite a few of these inventions adopting various combinations of simple machines fed down the centuries. Sticking with the cotton industry, the spinning jenny is one example of old technologies being used for increasingly mechanized processes, and played an important role as a precursor in the 18th century to a new industrial age—the ***age of the machines***.

Also from the Islamic world and key to the age of the machines came the invention of the crankshaft, for converting reciprocating motion into rotational motion; and the camshaft, for phasing the rise and fall of reciprocating devices in a timed or phased sequence. Both would be crucial in the design of the steam engine and for the ***reciprocating engine*** (internal combustion engine). They were each devised by Ismail al-Jazari (1136–1206) in 1206 and were widely used in time clocks and pumps throughout Mesopotamia.

The development of interleaving elements of simple machines such as these extended further, however, into the first programmable devices. Surprisingly, as with the invention of the organ, this advancement also came as a result of the desire to reproduce music. In the 9th century, the first music sequencer was described in *The Book of Knowledge of Ingenious Devices*, written by three Persian brothers—known as the Banū Mūsā—working in Baghdad, in modern-day Iraq. This written compilation describes an extraordinary range of devices, indicating the existence of an established catalog of simple machines manipulated so as to repeat, or replicate, spontaneous actions. They included the music sequencer, an early precursor to the pianola, which operates on rolls of perforated paper, or the recording disc of the 19th century.

Recorded by al-Jazari, a classic example of integrated machine technology displayed in this water pump involving rotational and linear gears.

The similarly titled *The Book of Knowledge of Ingenious Mechanical Devices*, written in 1206 by Ismail al-Jazari (who cited the Banū Mūsā's book as an influence), contains detailed drawings of a water-powered flute of extremely large proportions that was developed on a precursor concept postulated by the Banū Mūsā brothers. Al-Jazari also explained the workings of programmable robots operating on the ***automata principle***, whereby cyclical repeating patterns are made to occur in synchronism with a clock mechanism or a recoil spring, repetitively reset by a cam follower.

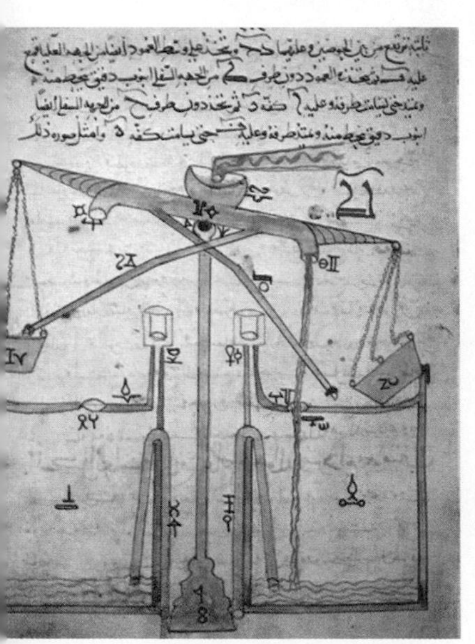

An application of the water-powered perpetual flute as recorded in The Book of Knowledge of Ingenious Mechanical Devices *by al-Jazari in 1206.*

Further evidence of creative talents of the post-Roman Arabic world in this complex and totally impractical elephant clock invented by al-Jazari.

CHINESE INVENTIONS

In China, where so many applications of simple machines originated, came the first escapement built into an astronomical clock tower. Developed by Su Song (1020–1101) in the 11th century, this invention preceded by two centuries the first appearance of such a mechanism in the West. It is considered to be the world's first ***analog "computer"*** since it operates on precisely the same principles as mechanical devices that were developed several centuries later. Su Song is also credited with designing and building the endless chain drive as a power transmitter.

Commonly, inventions outpaced discoveries, with the latter sometimes informing the practical application of the former. One such example is the powder rocket, which was developed from an observation of the chemical properties of gunpowder used as a propellant. Rumor surrounds the origin of the powder rocket, with some writers claiming that the Chinese were using rockets in the 10th century, although no firm evidence exists until the 13th century when there are substantiated reports of "fire-arrows" and "iron pots," reportedly audible for 15 miles (25 km) as they struck the ground and exploded. From early primitive devices emerged a wide range of projectiles powered by gunpowder, with treatises and manuals appearing in the 14th century containing detailed descriptions and explanations of several military devices. In part thanks to these written works, the technology spread to neighboring regions such as Laos and Korea, stimulating their eventual migration to eastern Europe via the Mongols, who adopted them in their expeditions westward.

WORKING MACHINES

So far, we have defined what we mean by "simple machines." ***Working machines***, some of which we have already learned about, are different only in that they compound several simple machines and integrate them into a homologated unit supporting a pre-defined task or objective. There were many of them, epitomized most commonly by the watermill and the windmill. This was because the Middle Ages saw a proliferation of such devices, particularly in England, where watermills and windmills became crucial to rural life, sustenance, and the origin of small business communities. Here, the proliferation of waterways and small-scale farming encouraged a level of autonomy and self-sufficiency upon which much of the English population thrived in relative comfort.

Then these devices were applied to the industrial production of materials such as fabric, leather goods, and paper, which could bring an income and broaden the opportunities for trade and exchange. This arose first in England

and Continental European countries and then was exported by way of the Crusades to the Middle East. Specifically, the windmill replaced the watermill. Both windmills and watermills applied the camshaft to the transmission of a rotational force, which was converted into directional energy. Wind power was more broadly applied in the Middle East than it was in Europe, simply because water, unlike wind, was scarce in many Middle Eastern locations. During the Middle Ages the migration of mechanical invention and discovery reversed and Europe began to emerge as the central focus of working machines.

THE ROLE OF MATERIALS

In the history of simple machines and working machines, ***materials*** were of such vital importance that they served as enablers, without which the empirical age would not have advanced as far as it did, let alone the scientific age, where learning would have no application without the fine-scale working of materials, notably metals. From this progression would come a renaissance in chemical science and the development and use of natural materials and substances for support and ancillary products for machines. These included greases, lubricants, belts, sealants, and artificial materials, such as rubber, nylon, and synthetic fuels or manufactured equivalent substances to replace or support natural products. But we are getting ahead of ourselves.

The Bronze Age

We discussed earlier that stone gave way to ***bronze*** tools—a shift that was of utmost importance; the Bronze Age is the first rung on a ladder of dating human progress and is no less a category than mechanical (and later electrical) engineering. It is marked by a cultural as well as a practical change in human capabilities, arriving in parallel with writing, unique styles of pottery, and a significant shift in the way societies and lifestyles functioned. The first shift to a Bronze Age society appeared in the Near East, embracing the Levant, Mesopotamia, Egypt, and India, around 3300 BC. But it did not arrive at a set chronological date everywhere; it is noticed in India from 3200 BC and in central and some parts of western Europe a century later. In Britain, the Bronze Age arrived around 2150 BC following a period in which ***copper*** was first adopted to find use in a wide range of utilitarian products, as well as items for personal adornment and decoration.

Defined by a specific form of pottery, the Bell Beaker culture arrived in the British Isles around 2000 BC, right at the end of the long period of Stonehenge development that had begun with the stone structure first being laid out in simplified form around 2900 BC. By mining ***tin*** and adding small quantities to

A precious commodity
Bronze could be worked into fine goods and decorative dishes, bowls, shields, daggers, swords, spearheads, and a wide range of esthetically pleasing artifacts. It provided a replacement workforce that had previously used antler picks to mine flint for tools that were applied to a wide range of tasks. Quite quickly, bronze also became a symbol of wealth, status, and power that would endure until the Iron Age. In a way, it was a currency of type to complement the age-old use of gold and silver, worked over centuries into fine goods and decorative embellishment and lauded as symbols of status. But neither gold nor silver had the hardness and durability—or the availability—to form the same universal application to simple or working machines that bronze had. An indication of the extent to which bronze dominated British culture was discovered in 1959 when the finest bronze hoard ever found was unearthed near the small town of Ely, outside Cambridge. In all, 6,500 pieces of worked and unworked bronze testified to the abundant use—and value—placed upon this precious commodity.

copper, the Beaker folk made bronze: a much stronger metal than copper and one that would dominate European metalworking industries for more than 1,500 years until the advent of the Iron Age. The smelting and refining skills introduced from the Continent came with successive waves of immigrants from the Low Countries and the Rhine. They brought with them the skills that would quickly make redundant the makers of stone tools that had endured for millennia.

Using bronze, communities could now fabricate tools for farming, for fashioning wood for houses and barn-like structures, and for weapons that would significantly expand the potential for warfare and tribal defenses. It is notable that the Beaker folk, who came from the Continent, were assimilated into the cultural heritage of the Ancient Britons rather than replacing them, and there is significant evidence to show that this new ***metal smelting process*** was beneficial to all. In fact, it was so beneficial that it caused the emergence of the first pan-European trade routes as essential components of large conurbations, which would in turn develop into small cities. And all this because south-west Britain had abundant supplies of tin for the smithies to work with. It was this that proved so profitable for the local population of post-Neolithic farming communities and the new "settlers" from the Continent, with their strong connections back to the mainland. Thus, the tin mines of what is today the English county of Cornwall became for centuries the dominant supplier of this essential element.

The entwined evolution of engineering and climate change
Largely outside the scope of this book, it must be said that human intervention in Earth's climate began with crop farming in the Neolithic period and is generally regarded as having started to influence the climate around 10,000 years ago, when through forensic analysis of ice cores and preserved biomass, the first increase in carbon dioxide levels is known to have begun directly as a result of human intervention. Later, we will have something to say about the ethical responsibility of mechanical and electrical engineering in the modern world, in which natural resources are being depleted and the modification of the environment is directly proportional to our application of engineering systems.

The Iron Age

Inevitably, the Bronze Age was replaced by the Iron Age, which took hold in the Near East around 1200 BC, a century later in central Europe, and in Britain by 800 BC. It grew out of a fascination with worked materials, the evolving knowledge about the properties of metals, and the skill of the metallurgist in working these different materials. Both bronze and iron required smelting, which called for vast quantities of charcoal. This was provided through widespread deforestation. Thus, the impact of human society on the environment really began to accelerate with the introduction of the Iron Age—a key enabling technology for industrial development and societal expansion.

Smelting

The earliest evidence of iron being recognized as a metal comes from around 3200 BC in Egypt, where small beads of meteoritic iron have been found in graves. Smelting was also used during the Bronze Age, but the conditions required for smelting iron were more demanding than those for gold, copper, or bronze. The fundamental development path for smelting technology is an implicit part of the development of mechanical engineering because without the sophisticated metals and iron products of the Industrial Revolution, none of the applications that underpin engineering today would have been possible. Mining sufficed to produce the metals employed for engineering and manufacture, but ***smelting*** was essential for producing the pure metals required for the new age of industrial-scale production.

Key to smelting were two essential controls: the temperature of the smelting furnace and the amount of air that could be pumped into the fire. To begin with, concertina bellows were used to raise the temperature of the fire and were capable of delivering 17.7 cu ft (500 l) of air per minute. This was sufficient for small-scale work but would eventually require water-driven bellows, which did not appear in Europe before the medieval period. The usual method was to use a charcoal fire to raise the temperature to about 1,830°F (1,000°C) for one hour, after which roasted ore and additional layers of charcoal were added. In the ensuing process, molten metal in a crucible would be reduced to droplets, the richest forming a slag, constituting 60–80 per cent of the total product as waste. This would be silica and iron oxides, which separately would rapidly choke the furnace, but react to form iron silicates with a much lower melting temperature, meaning they can run out of the furnace as tap slag. Sometimes, the ore would have the optimum mix of silica and iron minerals without any intervention, but if there was an excess of iron then a small quantity of iron oxide would be added. Conversely, if there was too much iron, a small quantity of silica (the flux) would be added.

More complex still was the smelting of ***silver*** from lead ores, which was labor-intensive since these generally contain a very small percentage of gold. Most sources are oxidized ores in the form of a hydrated lead carbonate known as cerussite. Alternatively, galena could be used for the extraction of silver.

Gold, by contrast, because it usually occurs as a metal, could be separated by crushing rock into tiny particles, a washing process being all that was necessary to separate the heavier metal. Yet as early as the Roman Empire, smithies were smelting auriferous iron pyrite, which contained gold particulates, almost down to the molecular level. Silica added to roasted iron pyrite would form a slag, which allowed the gold to separate and coalesce. Slag heaps from this process have been found at Roman mines in Portugal.

Despite the high value of silver and gold, it was the spread of iron in the 1st millennium BC that made and broke empires. Accordingly, throughout Europe and the Mediterranean world from the first millennium BC iron was smelted universally by what is known as the ***bloomer method***. The furnaces for this process were large and the general smelting conditions were not dissimilar to those described above for earlier processes. Produced as a solid, pasty mass, the iron bloom contained some proportion of slag, which was removed while still white-hot by a process of hammering out the residue to leave wrought iron, which could contain very small quantities of carbon. There are examples where sufficient quantities of carbon exist within the wrought iron for it to be classified as steel. Unbeknown to the smithies at the time, most bog ores (hydrated iron oxide minerals formed by groundwater flowing into wetlands) contain phosphorous, which leads to brittleness and today is discarded. Interestingly, in the ancient world, phosphoric iron was specifically selected (although the smithies only knew of the phosphorous by reaction of the smelted iron) for cutting tools such as knives and scythes. However, while advantageous for keeping its edge, such a blade was easily broken and the practice was discarded by the Middle Ages.

Chain mail
The essential applications of materials, chemistry by trial and error, skill, and an accumulating base of knowledge and information converged to produce the capability for moving to an industrial society built around mechanics and civil engineering. Some of the processes were learned due to demand, when form followed function. One example was the production of chain mail, the simplest examples of which contained several hundred yards of links. To produce these, a draw-plate technique was required, whereby a hardened metal has holes through which a softer metal is drawn. These links could only be produced by remorseless hammering and the draw-plate technique would have required a great deal of time and effort. But this was more readily sought due to the flexible nature of the finished product, which would more easily follow the contours of the human body.

Another form of metal, ***cast iron*** was first produced in China in the 6th century BC. It was discovered that if the temperature was increased and the conditions caused a reduction in the volume, carbon would likely dissolve in the iron as it formed, which reduced the melt temperature to 2,190°F (1,200°C). Under that condition, the liquid iron could be made to run out of the furnace with 3.5–4.5 per cent of carbon. With a part carbon content of 0.2–1 per cent, iron is known as steel. However, getting the quantity precisely correct, with the correct amount of carbon, was problematical. The ideal would have been to produce iron with 1 per cent of uniformly distributed carbon, achieved by mixing wrought iron with charcoal and wood inside sealed crucibles at a temperature of 2,730°F (1,500°C). This "crucible steel," forged but never cast, became available in the central and southern parts of Asia around 2,000 years ago.

The production of ***brass*** dates back to the 1st millennium BC. It is an alloy of copper and zinc but was difficult to produce. However, once the technology for its production had been mastered, it began to replace bronze. Despite the fact that it is more abundant than tin, the high volatility of zinc and the temperatures required to melt it (787°F/419.5°C) require precise control over the production process. Because of this, brass would not become available commercially in large quantities before AD 1000, primarily in India, from where it was exported, traded, and eventually adopted in Europe.

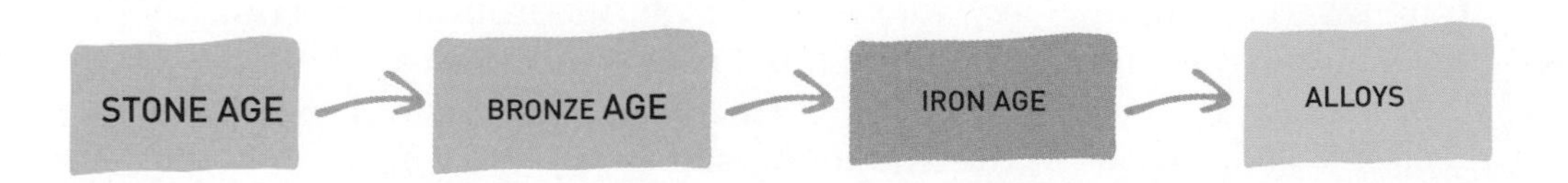

THE FORGING OF GLOBAL CONNECTIVITY

One of the greatest consequences of the Bronze Age was the explosion of connectivity that it brought across Britain and the European continent. From 1600 BC, vast networks of communication and social interactions were built, providing a pan-European economy, diffusing technology and access to materials, and flattening cultural enclaves into a homogenous continental sharing. Exchange of goods, roads, sea routes, and warrior networks were established as a result of the desire to access and acquire this vital resource. Thus began a 3,000-year process of globalization that would fuel the Industrial Revolution of the 18th century via an Age of Enlightenment.

Fired by the Bronze Age, a uniformity of access continued to course through the Mediterranean and European world long after bronze was replaced as the material of choice, connecting states, cities, and people and providing opportunities for an unprecedented exchange of knowledge to mutual advantage. And while the methods and practices for producing essential materials empowered the Industrial Revolution with new technology, great progress was made in coupling advancing capability to make complex machines even better. One example of this was the use of water power to replace the enormous (more than 3 ft 3 in/1 m in diameter and around 10 ft /several meters high) manually operated bellows that had hitherto been used for smelting metals. With this capability, a smelting operation could be maintained for several weeks at a time using only a handful of highly skilled men. This provided a powerful incentive to move beyond the tools and techniques inherited over preceding centuries and to harness a new age of scientific and mathematical knowledge that manifested itself as sophisticated machines built on principles of mechanical engineering.

Machines of war reached new heights of sophistication in medieval times at the peak of the empirical age, seen here in this depiction dated 1470, showing guns.

Chapter Two

UNDERSTANDING FORCES—THE SCIENTIFIC AGE

The Driving Force of Progress—Forces in Action—The Rise of the Machines—Analytical Machines—Newton's Laws in Action—Putting Principles into Practice

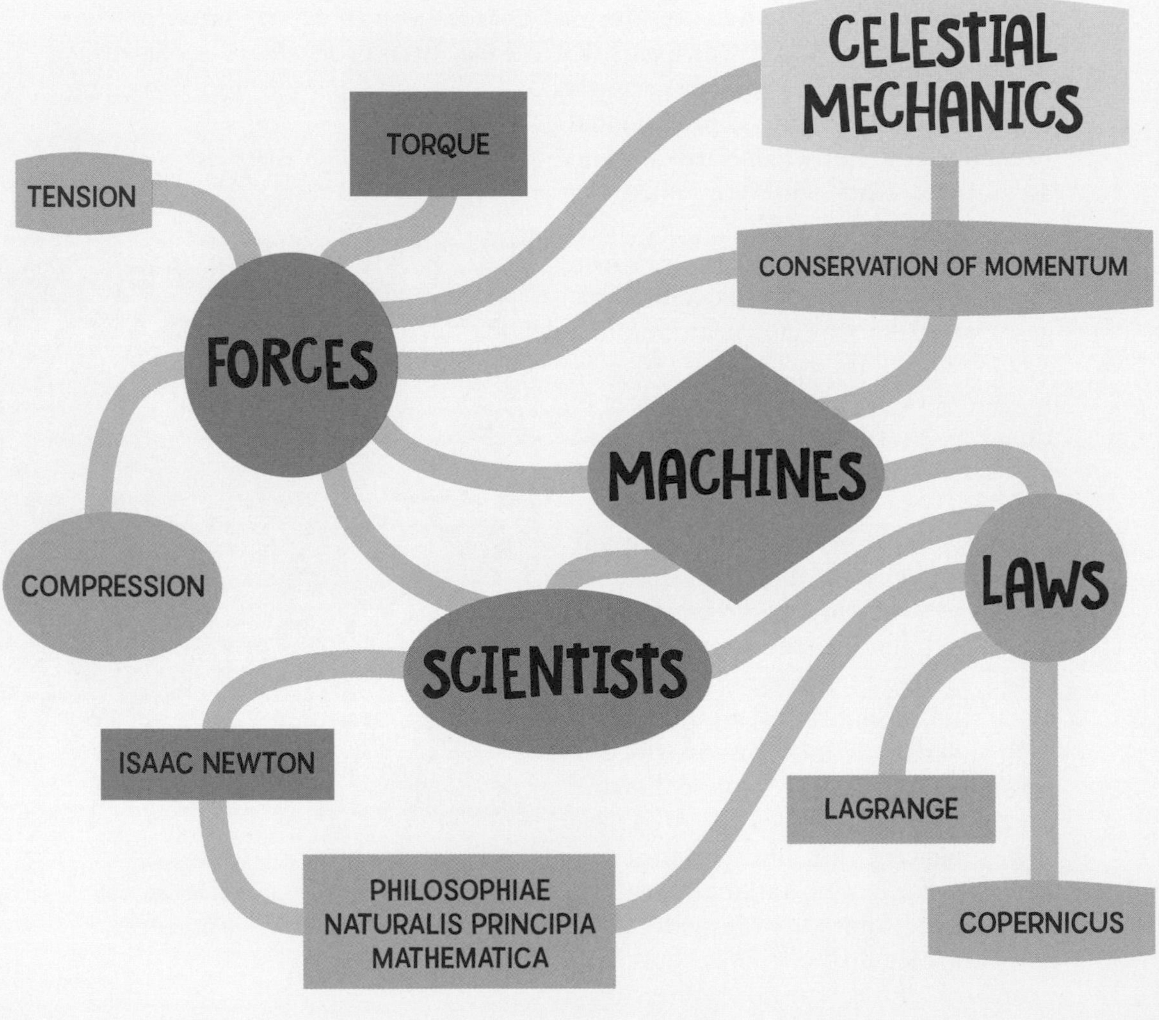

CELESTIAL MECHANICS
TORQUE
TENSION
CONSERVATION OF MOMENTUM
FORCES
MACHINES
LAWS
COMPRESSION
SCIENTISTS
ISAAC NEWTON
LAGRANGE
PHILOSOPHIAE NATURALIS PRINCIPIA MATHEMATICA
COPERNICUS

THE DRIVING FORCE OF PROGRESS

It is almost axiomatic that humankind has achieved the greatest technological leaps, involving many aspects of different mechanical principles, in times of conflict and war. Indeed, before the scientific age, the classical world was defined by this. In giving brief mention of those capabilities, introduced largely during the Classical Age (8th century BC–5th century AD) but improved upon during the medieval period, we move from science that enabled the evolution of complex machines to ***engines*** that used ***forces***, the types of which comprise the substance of this chapter.

The societies that grew out of the development of both simple and complex machines through the Bronze Age and the Iron Age went on to frame the city-states and nation-states that competed with each other in terms of trade, commerce, power, and influence. In turn, these stimulated experimentation, trials and inevitable errors, until sophisticated capabilities evolved utilizing basic materials but innovative and ingenious ways of making things, and in so doing underpinning the ***age of mechanics***. And of such stuff were the engines of war designed and fabricated—using basic mathematical principles, some defined and some as yet unwritten, but acquired by application through the empirical process.

During the Classical Age, high-earning craftsmen and skilled engineers were sought out and employed by powerful men and women for designing, perfecting, and building all manner of devices for defense and attack. One of the earliest was the bow and arrow, which, its draw strength limited by the 44 lb (20 kg) pull capacity of an archer's arm, had limited application. Several mechanical devices were also developed for surgical operations, one by Hippocrates, using a winch to apply torsion on implements designed to reset bones. And it was here that the principles of ***torsion, tension, compression forces,*** and ***shear loading*** were paramount. Before we explain why, let's look in a little more detail at those forces.

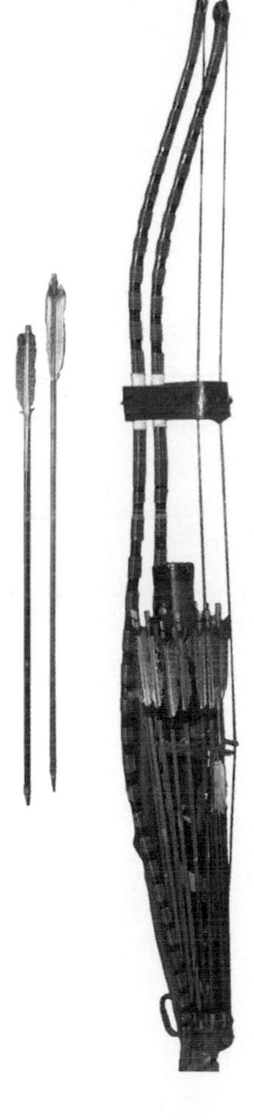

The application of torsion, tension, and compression began with the adoption of principles fundamental to mechanical engineering, beginning with the bow and arrow, here represented by a Japanese war bow.

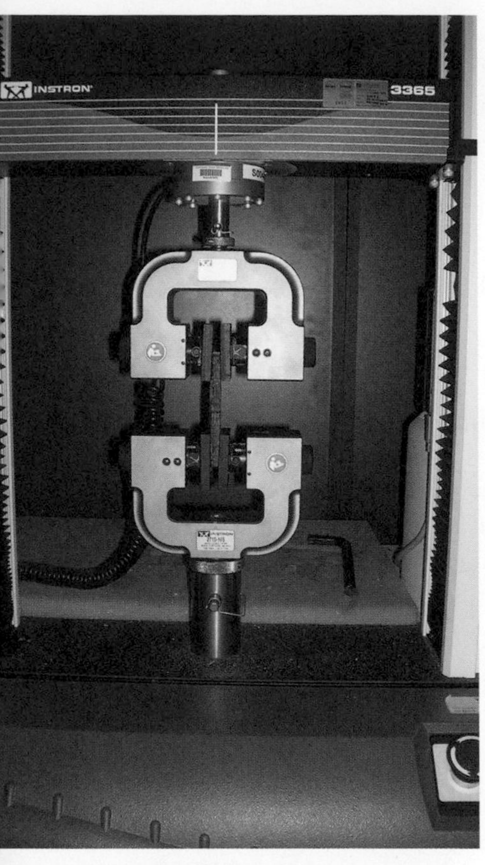

Tensile strength demonstrated on a coir composite by pulling it, stretching the material until it fractures.

Torsion

Getting its name from the Latin *torquere*, meaning "to twist," torsion is created through the act of twisting two ends of a material or structure in opposing directions, much as a cloth is wrung out by hand to remove water. This deformation is measured as the ***shear force*** acting tangentially to a surface, and can be an actual surface or the defined surface within a structure, such as a beam. It is measured as a load factor parallel to the surface on which it operates.

Torque

Torque is an essential component of all machines up to and into the age of mechanical engineering and to the present day. Defined, it is the application of a ***rotational force***, the equivalent of a linear force, and its derivation goes right back to Archimedes and his use of levers.

Tension

In an engineering application, tension (from the Latin *tensio*, which means "constriction") is the process of stretching, and is measured as the balancing force causing ***extension*** (hence ex-tension). In mechanics, it implies a force transmitted axially, either in a cable or a single-string rope, or at each end of a truss structure or even in a three-dimensional object.

Strictly, in physical expression, tension is a transmitted force that's sometimes referred to as a ***restoring force***, with the magnitude of that force measured in Newtons. In mechanical structures, tension can be used to describe the amount of elongation and the applied load that will cause failure—a factor dependent on the force per cross-sectional area and not just the force alone. Tensile force per area is measured as to the tensor, which is itself a measure of the stress applied.

Compression

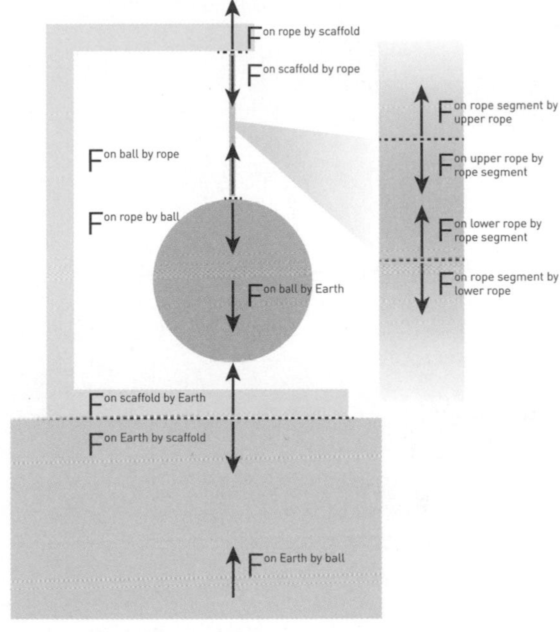

A figure depicting the forces involved in suspending a ball from a scaffold by a rope. Each force is shown at its point of action and is labeled by the object it acts upon and the object it is produced by. The tension in the rope is shown as it acts on the ball and scaffold, and also in a segment of the rope.

Compression is a force that's crucial to understanding the connection between complex machines and engines. In fact, it's arguably the most important consideration when dealing with possibilities raised by fluids and gases, which were to become critical in powering the world from the 18th century onwards. Compression is defined as the application of inward pressure from uniform points at the extremities of a fluid (be it gas or liquid) or the pressure applied, uniformly or tangentially, to a material object. It can take many forms and produce very different results. In a solid object, compression can take place along some direction, while other parts may be under traction—a force applied to generate motion between a body and a tangential surface (a measure of which is defined as the maximum force between a body and a surface).

The consequence of placing a material under compression is that it will be ***deformed***. In physical terms, that can be visually apparent or it can be indiscernible to all but the most sensitive measuring devices, but in all cases it causes the atoms and molecules to change, potentially modifying the tensile or torsional strength of the material. It is the Rubik's Cube of force integration; change one and you change another! And that influences the way the material behaves. In some cases, deforming a material can cause a reactive force to counter the deformation and achieve for the material a level of stability. Reactive forces, meanwhile, can cancel the potentially damaging effect of compression. Of course, we will encounter compression in fluids when we come to the internal combustion engine, the effects of which liberate stored energy as a reactive force, and this will also be applicable to some forms of rocket motor, too.

Compression test using an Instron universal testing machine to measure the compressive strength of a composite cylinder.

Shear loading

Shear loading is measured in mechanical engineering when two connected components are subjected to equal but opposite compression (***tensile***) forces, and can also be applied to fasteners subject to a value known as shear loading—in the case, for instance, of two plates secured by a bolt or a rivet. This is calculated by a bending moment given as F(t1 + t2) where F is the load and 2t1 and 2t2 are the thickness of the two plates secured together by the fixture.

FORCES IN ACTION

Now back to Hippocrates and resetting bones through the application of torsion. Designed to realign limbs in the patient, the windlass was used for stretching the human body for surgical practice (it also had an unhappy function for torturing prisoners), and consisted of a winch with handspikes that could be attached to a bench for visiting surgeons, so that the surgeon could stretch the body. Improvements employed a screw thread instead of a winch, and a still later version turned a screw by engaging a gear wheel on the winch. The definitive version used pulleys and sometimes geared winches to gain mechanical advantage, and it was from this that the next major step in the mechanics of war machines began.

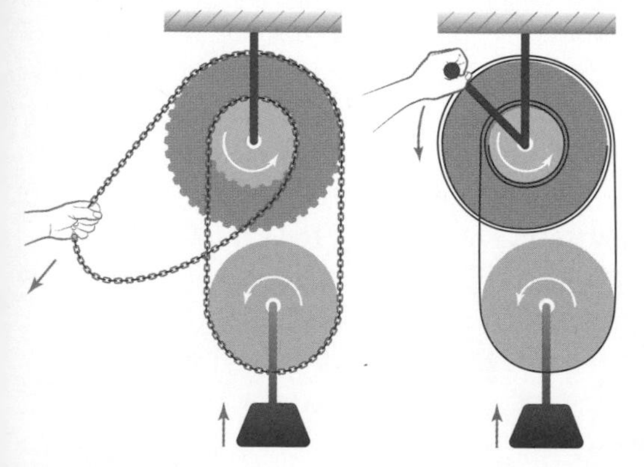

Comparison of a differential pulley or chain hoist (left) and a differential windlass or Chinese windlass (right). The rope of the windlass is depicted as spirals for clarity, but is more likely helices with axes perpendicular to the image.

Engines of war

Dionysius of Syracuse is attributed with the invention of the catapult. This device, for the first time, enabled the capture and release of a greater amount of energy than could be stored and released by an archer, which, as described earlier, was limited by the muscle power available in the upper limbs. The catapult resembled a huge crossbow and had a large bow attached to a stock supporting a slider where the bolt was held in a groove. The rear end of the slider engaged with the bowstring so that before the bow was spanned, the slider projected far beyond the end of the stock. A crosspiece at the end of the stock was shaped so that it could be engaged with the stomach of the archer, allowing him to span the weapon by leaning the stock against a fixed surface, such as a wall or an immovable rock, and pushing against the stock with his body. The trigger took the form of an iron bar with a claw at one end that was hooked over the bowstring. The instrument was fired by pulling a hinged bar from beneath the back of the trigger so that the claw would tilt up and release the bowstring. From this basic application of forces, much larger devices were developed, including one mounted to a fixed base and spanned by a windlass. Others could shoot two bolts at the same time.

From this simple war machine grew an ever-expanding inventory of complex machines and what are rightfully described as "engines of war," through development of individual inventions into combined and compounded capabilities based on the application of tension, torsion, compression, and torque.

THE RISE OF THE MACHINES

We have considered simple and complex machines and the emergence of an age of war machines. All of these operated utilizing principles that are the foundations of mechanical engineering. But to bring us to the Industrial Revolution and the Age of Enlightenment in science and mathematics, it is necessary to appreciate the shift towards a new set of definitions: ***mechanical principles, thermodynamics, materials, structural analysis,*** and ***electrical energy***—both in its production and application. NB: thermodynamics will be dealt with in the next chapter and later we will embrace the new age of engineering design using computer-aided design (CAD) and computer-aided manufacturing (CAM), frequently unified as CAD/CAM. It is this combination of fields that characterizes mechanical engineering in the 21st century.

Newtonian mechanics

Newtonian mechanics is sometimes referred to as "***classical mechanics***," which explains the world of motion, how motion can be applied through the use of energy or force, and how it can be predicted to operate in the future or to have performed in the past.

If we accept the more modern expression of classical mechanics, we define the laws that govern natural forces. These underpin all of mechanical engineering as we express it today. As such, it is fundamental to understanding how things work but—arguably more importantly—*why* they work in the way they do. Understanding this provided a transition from the empirical to the scientific method. It should be noted that it was not defined by Newton alone, but rather by a group of thinkers in the 17th century, each working off the writings of the other, but together they provide a frame within which mechanical engineering could develop.

What is mechanics?

If we break it down into its constituent parts, ***mechanics*** describes the movement of large objects, and this can extend to fields of science, including astronomy and space travel. It also underlies the essential principles upon which mechanical engineering is based. It is established in large part on the mathematical models defined by Isaac Newton (1643–1727), a 17th-century polymath of outstanding ability and creative genius. This period also saw integration among intellectual thinkers, coming together to solve universal problems. For instance, much of Newton's work on celestial mechanics and gravitation was based on the earlier observations and writings of Johannes Kepler (1571–1630) who, along with Galileo Galilei (1564–1642), defined a substantive challenge to orthodoxy as proclaimed by the Church of Rome. Kepler in turn defined laws of planetary motion from the observational work of Tycho Brahe (1546–1601). The bridge between these two generations established a philosophical foundation upon which the modern world is built, ensuring a rational explanation for natural forces based on observation, measurement, and calculation.

Sir Isaac Newton, painted by Sir Godfrey Kneller.

While we may regard this work as a display of the ***scientific method***, in fact Newton always regarded it as the empirical approach through what he described as "inductive reasoning." It is this approach that defines what lay at the foundation of Newton's great work, *Philosophiae Naturalis Principia Mathematica*. The concept is a way of thinking and is fundamental as a process in deduction—which is not wholly empirical but is conducive to reason based on findings through observation and experimentation. Any good engineer needs to master this structured process to extract applications from both absolutes and constants in physical law and in mathematical models.

Newton's laws of motion

Newton's laws of motion are fundamental to the definition of the world in which engineering operates, and as such do not directly relate to a specific object or machine but rather to laws governing the motion of all things.

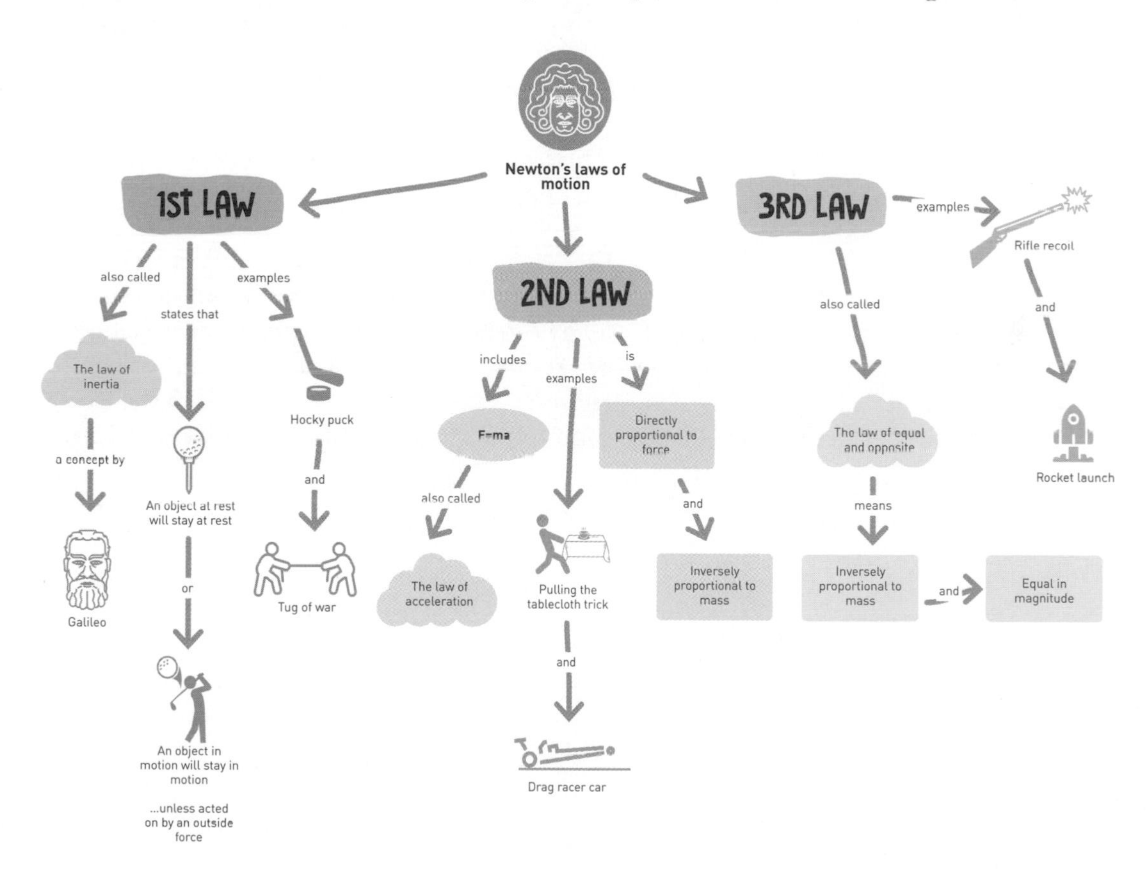

Newton's first law of motion—inertia

His first law (for which there is no formula) states that a body with a measurable mass will remain at rest or in motion unless a force is brought to bear upon it. This is the classic law of inertia and harks back to Galileo, replacing the empirical concept of motion requiring a cause; in other words, that a body requires a continuous force applied to it to keep it in motion. The fact that Galileo's theory is not so is reflected in the motion of a planetary body (such as Earth or the Moon) through space or the movement of an artificial satellite around the Earth. Once that motion has been produced (by a force), the motion will continue until another force is applied to change that motion—either accelerating or decelerating it. That force can be generated by internally or externally applied force. For instance, if a satellite has a rocket motor capable of applying a force, either to speed it up or slow it down, and that force is applied, it will change the motion of the satellite. Should that satellite run into transient chemical molecules, such as might be encountered at the fringes of the atmosphere, it will slow down through friction.

Newton's second law of motion—acceleration

The second law stipulates that the rate of change in the momentum of a body is directly proportional to the force applied. But this is only so for bodies of constant mass, and in this context a net force is equal to the rate of change of the momentum. This defines the conservation of momentum, and any variation in the mass of the object will cause a change in momentum that is not the result of an external force. The way Newton described this law was typical of the expressions of the day. Today, the law is expressed as: the change in momentum of a body is proportional to the impulse impressed on the body and occurs along the straight line upon which the impulse is discharged.

Second law formula

Assuming that the mass (m) is unchanging, Newton's second law is written as $F = ma$, showing that force (F) equals mass times acceleration (a). It follows that if no acceleration is applied to the body, no net force is acting upon it.

Newton's third law of motion—equal and opposite

The third law states that forces between two objects are of equal magnitude and opposite direction. This binds forces as being interactive between bodies, or in different elements of a finite body, so that there is no condition in which there is no reciprocal action and reaction. This is probably the most intuitive law in that it defines the way two surfaces relate to each other. For instance, when an aircraft is taking off, the wheel tyres push against the runway and the runway pushes against the tyres. This law is arguably more important for the principle of conservation of momentum than it is for the measure of action and reaction.

> If object A presents a force F_A on a second object B, B simultaneously exerts a force F_B on A; the two forces are equal in magnitude and opposite in direction, $F_A = -F_B$.

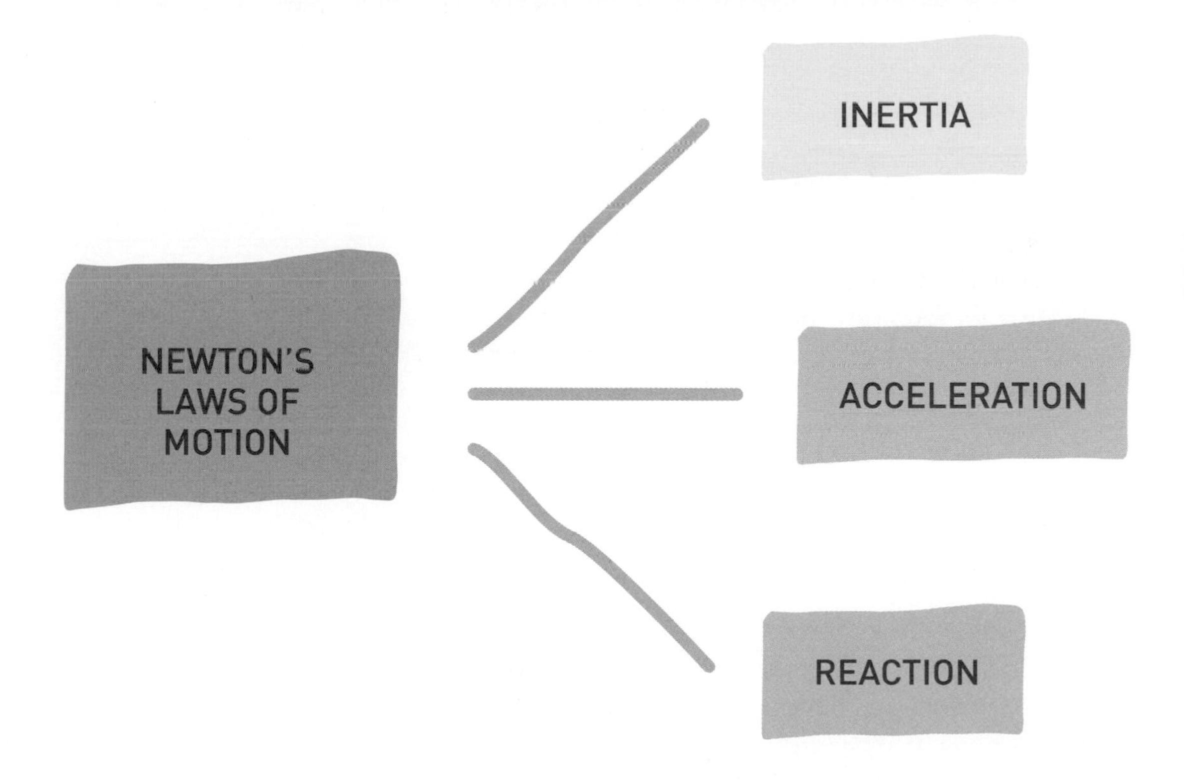

ANALYTICAL MECHANICS

One of the early springboards to a more nuanced view and interpretation of the physical laws that would power the Industrial Revolution was that of ***analytical mechanics***. It emerged during the 18th century and diverged from Newton's view of ***vectors***—motion, momentum, acceleration, etc—and entered the realm of ***scalar properties***, that is, laws governing the whole or integrated system which involves kinetic and potential energy. In embracing the wide spectrum of forces, this simplifies the modeling of integrated forces and makes it a lot easier to solve a wide range of mechanical problems. Two aspects are worthy of consideration, with examples in real-world applications.

One of these is analytical mechanics, which also defines the co-ordinate system, or frame relative to a reference. These are usually set to provide separate, independent co-ordinates to simplify Lagrangian equations of motion. Joseph-Louis Lagrange (1736–1813) was a noted astronomer and mathematician most famous for defining analytical mechanics, and in several ways he advanced the ideas of Newton to another level. His work set in train a development of mathematical physics for the 19th century. Arguably his greatest contribution was the ***calculus of variations***, which uses the tiny variations in functions and functionals to calculate the maxima and minima of the functionals. This enables a calculation to find the relative location of the functionals to real numbers. But we are going off at a tangent. Back to the more direct and practical way Lagrange defined the way we view forces in play today. One example is classical ***Lagrangian thinking***.

Lagrangian points

In attempting to solve a three-body problem, taking the initial positions and velocities of three point masses and determining their subsequent motion according to Newton's laws of motion and of universal gravitation, Lagrange provided the basis upon which it is possible to take the Sun, the Earth, and the Moon and show areas of neutral gravitational attraction. These are known as Lagrangian points—places in the three-body system where a separate body will remain, in large part, motionless with respect to the other three.

To give an example, there are five Lagrangian points of practical use in today's Space Age—five places where spacecraft or structures can be placed so that they retain their relative position without any need for an action or reaction. The L1 point lies on a line defining the center of two masses: M1 defining the Sun and M2 defining the Earth, where the gravitational pull from M2 partially cancels out the gravitational pull from M1. The L1 point is on the solar side of the Earth and because it is on an inner track, an object placed there may be considered to have a higher relative speed (to remain in orbit). However, the Earth's gravity cancels out some of the pull from the Sun and increases the orbital period by seeming to slow it down. The L1 point in the Earth–Sun system is approximately 1 million miles (1.6 million km) toward the Sun and remains fixed at that distance synchronous with the orbital period of the Earth—hence it appears to remain in one place.

The L1 point has been used for scientific satellites by both the US and Europe since the late 1970s. Several additional proposals have been made for the practical use of the L1 point, not least a solar observatory placed at this location, where it will forever remain upwind of the solar particles streaming from the Sun (known as the "solar wind"). The highly energetic streams of charged particles can cause disruption to communications satellites at times of intense solar activity. These applications are directly attributed to the work of Joseph-Louis Lagrange, but in fact the first three were discovered by Leonhard Euler (1707–83) a little time before.

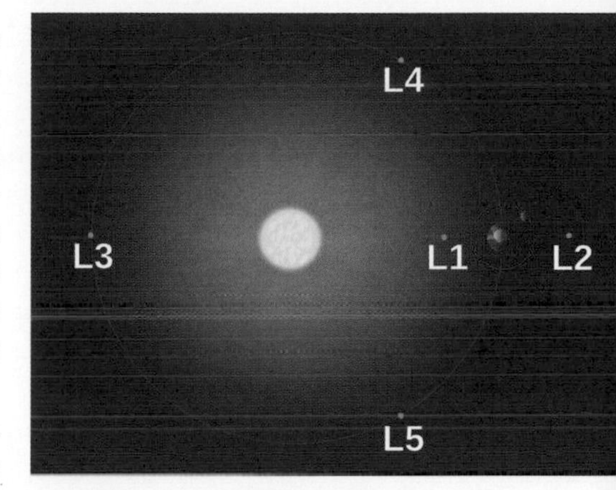

Lagrange points in the Sun–Earth system (not to scale) wherein a small object at any one of the five points will hold its relative position.

NEWTON'S LAWS IN ACTION

An important product of Newton's second law of motion states that the momentum of a body is equal to the magnitude and the direction of force imposed upon it. This is a vector quantity, as is ***velocity***—designating speed plus direction and not speed alone (as is frequently assumed). It determines that a force applied to a body can change the magnitude of the momentum and its direction. This law is one of the most important when it comes to designing and building machines.

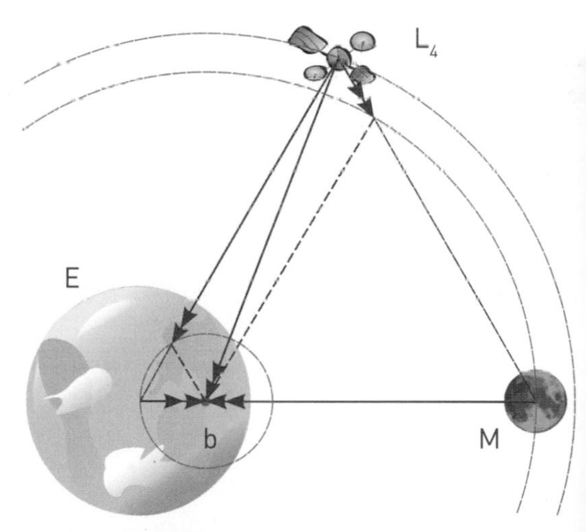

Diagram of stability of a satellite at L4 Lagrange point with b the barycenter of the Earth–Moon system and red arrows showing the origin and direction of forces of acceleration angular velocities.

By application, Newton's third law is one of the most crucial for realizing the dream of humankind to travel through space. Because it states that every action causes an equivalent reaction equal in magnitude and opposite in direction, this is the law that allows a rocket motor to operate in a vacuum and to propel a space vehicle in the desired direction. In the case of the action, a sustained process of chemical combustion causes an expansion of resulting gases that, if allowed to escape through a directional orifice, such as a jet exhaust, will cause an equal and opposite reaction. As a consequence, the body of the combustion chamber, if attached to another body, will cause the assembly to move in the opposite direction to the vector of the expelled gases.

This is the principle upon which the rocket motor works and whether missile or spacecraft it seems an identical function. It is of course the principle of the jet engine, too, but in the case of the rocket motor, the fuel and the oxidizer essential for chemical combustion are carried integrally within the body of the projectile. In a jet engine, by contrast, only the fuel is carried internally, the oxidizer (oxygen) being taken in as the jet aircraft flies. Furthermore, the inward flow of oxygen to start the jet engine from a rest position is accelerated (as though the engine were moving) by a compressor, which sucks in the oxygen required for combustion.

Newton's fourth law?

There is also the presumption that in his work, Newton set down the principles upon which it could be said a fourth law exists: that of the ***principle of superposition*** in which forces can be added together like vectors. This means that two separate causations, when brought to interplay, produce a response that is the sum product of the two induced by separate source causations. It implies that two wave forms, when originating from separate point locations, can each set off and propagate without knowledge of the other but that when they interact, the product of that causes a separate vector.

Principle of superposition formula

If an input designated A indicates a response X and a second input B produces response Y then (A + B) produces a new response (X + Y).

When written, a function that demonstrates this is known as a ***linear function*** and the superposition principle can be known by two properties, one ***additive***, the other ***homogeneous***:

- additive: F(x1 + x2) = F(x1) + F(x2)
- homogeneous: F(ax) = aF(x) for scalar a

This is not merely academic. There are many applications in engineering where the superposition principle is modeled into linear systems in situations, for example where the load on a beam is the input stimulus and the output stimulus is the deflection of the beam: two wave forms superimposed to create an additive reaction. From this, several mathematical techniques—such as Fourier's law of heat conduction, linear operator theory, and the Laplace transform—come into play.

PUTTING PRINCIPLES INTO PRACTICE

In a period of less than 150 years—from the argument by Nicolaus Copernicus that the planets went around the Sun and the publication by Newton of his *Philosophiae Naturalis Principia Mathematica*—the scientific age was born. From this would come a new dynamic: the commercial world of industrial development and manufacturing techniques, dependent on the principles and forces laid down in the 18th century. But these in turn would stimulate experiment, trial and error, and the principles of thermodynamics, which, when coupled with those discoveries outlined in this chapter, would create a new world of invention and discovery.

Chapter Three

MANIPULATION OF FORCES

Harnessing Heat and Power—Steam Power—James Watt and Thermodynamics—Defining Thermodynamics—Accelerated Progress

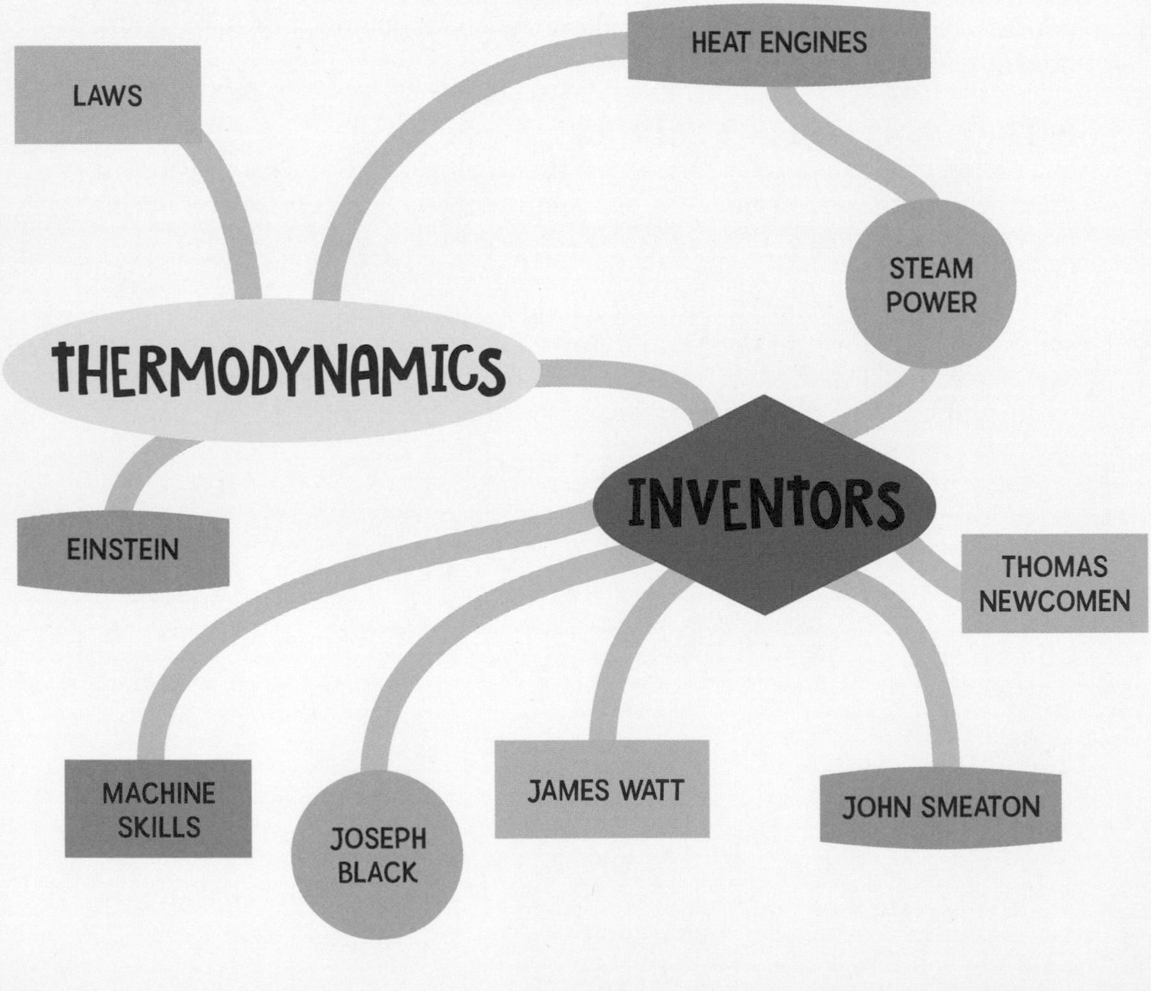

THERMODYNAMICS

LAWS

HEAT ENGINES

STEAM POWER

EINSTEIN

INVENTORS

THOMAS NEWCOMEN

MACHINE SKILLS

JOSEPH BLACK

JAMES WATT

JOHN SMEATON

HARNESSING HEAT AND POWER

So far, we have looked at empirical thinking as it relates to the fabrication of tools, compensating devices, and simple machines. These made life easier, influenced the evolving structure of society, and created static communities from disparate hunter-gatherer groups. The desire to take hold of the Earth's plentiful supply of riches and fashion them into instruments of change began the farming revolution 10,000 years ago. From that transformation, of nomadic wandering humans into settled groups bringing food to the family rather than taking the family to the food, came the baseline from which the long history of mechanical engineering began—directly in the case of farming, with instruments and machines for efficient and time-saving changes.

But it was not enough to know the forces and the physical laws by which these improvements were made. As simple devices were developed into complex and advanced machines, increasing use of heat, light, and chemical compounds to take engineering to another level provided the principles upon which the Industrial Revolution was developed and expanded. By the 18th and 19th centuries, complex machines operating at the peak of basic mechanics—weaving looms, cotton mills, steel production, and factories producing everything from pots and pans to bicycles and boats—reached the limits of known technology, which was the fundamental application of mechanical engineering.

The essentials of thermodynamics

To understand the way in which materials could be manipulated, it is necessary to understand how the four laws of thermodynamics underwrite the way heat, work, and temperature are related to energy, radiation, and the properties of matter. Thermodynamics is a branch of physics that embraces both science and engineering through physical chemistry as well as chemical engineering and mechanical engineering. This category applies to almost every aspect of the field and includes scientific studies outside the zone of engineering; its principles and applications are also involved in climate research and general meteorology, for instance.

The word "thermodynamics" originated as a hyphenated link between "thermal" and "dynamic," then it was accepted as the noun "thermo-dynamics" before being collectivized into the term "thermodynamics" in 1840 as claimed by the American Donald Haynie. His noted 2001 textbook *Biological Thermodynamics* defined the associated studies of heat and power that best describe what thermodynamics is about: a set of generalities that operate as principles before application into laws.

In a historical context, the search for a unification of heat and power was begun by Otto von Guericke (1602–86), a German scientist who gave his name to a wide range of studies and inventions, including defining the physics of a vacuum, describing electrostatic repulsion, and interpreting the laws of nature as expressions of a divinity. In 1650, Guericke had designed the first vacuum pump—after expressing its state in scientific terms—having been inspired by Aristotle's assertion that "nature abhors a vacuum." Just a few years later, the British natural philosopher and polymath Robert Hooke (1635–1703) built an air pump that was used to test the relationship between pressure, temperature, and volume. From these experiments, Robert Boyle (1627–91)formulated his eponymous law that pressure and volume are inversely proportional, if the temperature and the quantity of gas within the closed system remain the same.

But it was in a different place that the real applications were made which would provide engineering with a boost in conceptualized design that would significantly increase the potential of existing machines to carry out more productive work. And to understand that, it helps to go back a few steps and look at the empowering innovation of the 17th century: the steam engine. For it was in 1679 that Denis Papin (1647–1713) laid out the working principle of the ***piston and cylinder engine***. However, being a theoretician, Papin failed to make the most of his concept, and it was down to Thomas Savery (1650–1715) to build an engine based on this operating principle. Nobody could have known that this would be the turning point in mechanical engineering that would define a new age of industrialization. But first, some background against which to place these momentous steps.

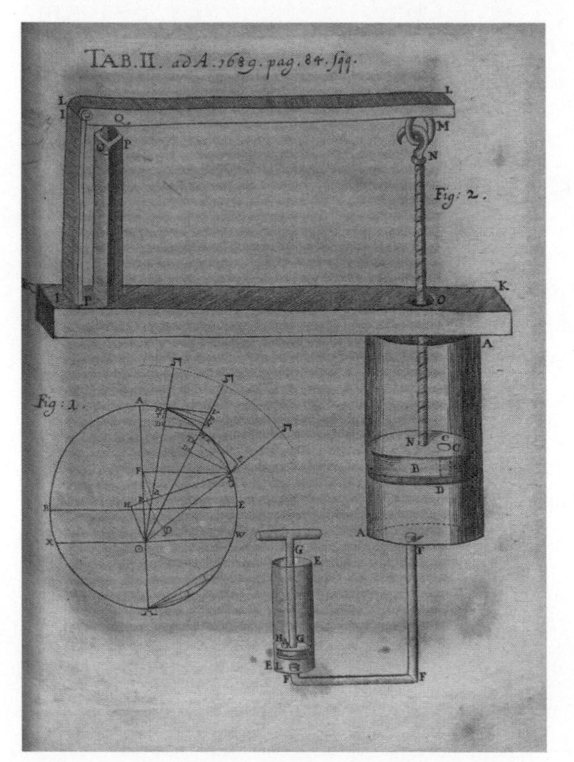

Denis Papin's experiments with the first steam engine displayed through an illustration from the Descriptio torcularis *published in 1689.*

STEAM POWER

Before the 18th century, industries were local and depended almost exclusively on the power of wind and water, with animals and men supplementing for additional energy in smaller activities. Watermills, windmills, and sailing ships were prevalent and still without the benefits of the scientific method. It was still the empirical age, with basic principles of operation little different from those used in the preceding 2,000 years. Nevertheless, the use of steam was inevitable once the initial experiments had demonstrated the practicality and the added leverage that steam could provide. It is perhaps somewhat surprising, then, to learn that the first experiments into steam power were made by people studying the atmosphere—the early stages of the first application of the scientific era into meteorology.

We have seen that it was the search for a vacuum that spurred the invention of steam power, and that came about because the ability to condense a small portion of the atmosphere trapped in a cylinder was a highly convenient way to achieve the desired vacuum. In harking back to Hero's aeolipile, Papin took the logic a stage further, but it was in demonstrating a Newtonian principle of action and reaction that Hero's device displayed characteristics of two enabling principles, the latter within thermodynamics, as we shall see. But there were some beneficial attachments to Papin's invention—based on the premise that by creating a vacuum beneath a piston it would be driven down by atmospheric pressure at a rate of depression dependent on the pressure of the atmosphere—including Papin proposing to the Royal Society that the device could be used to pump water out of mines.

Atmospheric engines

The general expansion of trade and commerce in the 17th century led to increasing pressure on the ability of mine owners to deliver raw materials extracted from increasingly deep shaft mines. There was a serious problem with flooding, especially in the lucrative Cornish tin mines that had begun to replace mining of alluvial deposits two centuries earlier. Earlier solutions using pumps powered by horse gins or waterwheels were overwhelmed with increasing quantities of water as shafts went deeper. The solution was pursued by Thomas Savery (1650–1715), who invented the fire engine, a device with steam in a cylinder condensed so as to draw water up a pipe past a non-return valve; more steam under pressure then expelled water past another non-return valve up a further pipe. Or at least that was the idea....

In practice, problems arose when it was discovered that the lower pipe could be no more than 28 ft (8.53 m) in length, and that meant that the entire apparatus had to be lowered halfway down the mine shaft in order to reach the level of the water. But the real limiting factor was that the extant technology could not come up with a boiler that was able to withstand the steam pressure required. In consequence, although protected by a patent issued in 1698, Savery's device saw little use and not a lot was heard of it afterwards. However, an associate of Savery, Thomas Newcomen (1664–1729) of Dartmouth, Devon, designed a slightly different device, which unknowingly harked back to the principle laid down by Papin. Newcomen worked within Savery's patent design, and when he inherited the patent on Savery's death in 1715, he took part-ownership of it and began to make improvements—after a long journey lasting several years involving grueling work, trial, and improvement.

Through several important business contacts, Newcomen had received help to set up a demonstration atmospheric pump system at the Earl of Dudley's Conygree Coalworks at Tipton in 1712 and it proved to be a

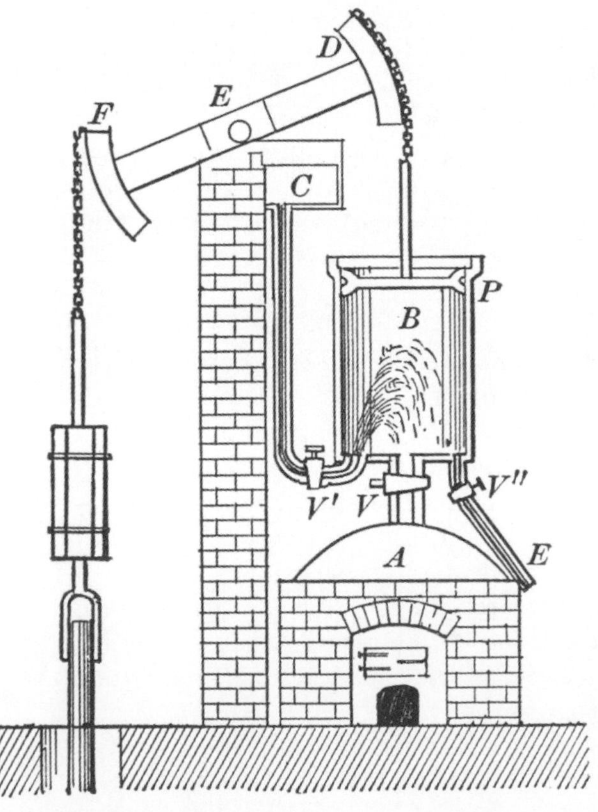

The Newcomen steam engine established a benchmark for a patent design that would herald a new age in industrial machinery. In this diagram, the copper boiler (A) is set above the cylinder, which produced low-pressure steam (at 0.07–0.14 bar), the action transmitted through a rocking beam, the fulcrum E resting on a solid end-gable wall of the engine house with the pump projecting outside the building. Pump heads were slung from a chain at arch head F while the in-house arch head D carried a suspended piston P working in cylinder B, the bottom end of which was closed, apart from a small admission pipe linking the cylinder to the boiler. Water tank C fed a small in-house pump slung from the smaller arch head, the header tank supplying cold water under pressure via a standpipe for condensing steam in the cylinder. At each top stroke of the piston, warm sealing water overflowed down two pipes, one to the in-house well, the other to feed the boiler by gravity.

DENIS PAPIN'S STEAM ENGINE

THOMAS SAVERY'S FIRE ENGINE

THOMAS NEWCOMEN'S ATMOSPHERIC PUMP

JAMES WATT'S UNIVERSAL APPLICATIONS OF THE STEAM ENGINE

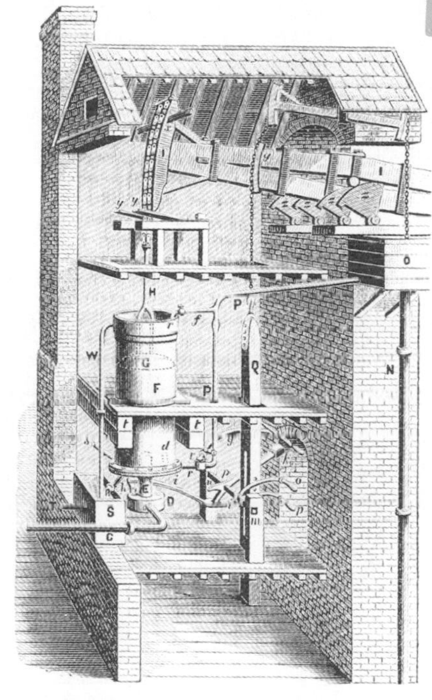

This pencil sketch of a Newcomen engine as modified and improved by Smeaton in 1775 was published in Popular Science 1877/78. Smeaton would come to be known primarily as the first civil engineer rather than for his work to improve the Newcomen design of steam engine.

winner—at least in comparison to what was around at the time. Although based on very simple principles, the engine itself was complicated, the working parameters being identified in an accompanying illustration. Despite this, the design was taken up and applied to at least 75 locations when Savery's patent expired in 1733.

The erection and assembly of Newcomen's atmospheric engines thereafter bred a new generation of engineers working not exclusively in the practiced arts of bridge building, farm machinery, and boats, but in the practical development of steam as a motive power—for static and mobile machines.

Improvements to Newcomen's design were inevitable. Forty years after Newcomen's death in 1729, John Smeaton (1724–92) conducted trials and experiments to improve the efficiency of the atmospheric engine and some progress was made, with improvements to parts and in the general design of the assembly.

The world's first mechanical engineer
Intended for a career in law, John Smeaton was attracted to the fine workings of small instruments, involvement with which allowed him to set up his own business, from which ensued the Eddystone Lighthouse off Plymouth, the third on site but the first to be completely successful. However, Smeaton would not be known for either this or his work on the Newcomen engine. Instead, he became the world's first civil engineer, distinguished by category from the profession of military engineering, and thus he is regarded as the first "mechanical engineer," as we would refer to him today.

Reassembled and on display in the Science Museum, London, the garret workshop of James Watt faithfully depicted as it was in the late 18th century. Credit Frankie Roberto.

JAMES WATT AND THERMODYNAMICS

The importance of the atmospheric engine, powered by steam, motivated industrious businessmen to seek a more efficient and more powerful system than the one that held sway for much of the 18th century. It fell to another Scottish inventor, James Watt (1736–1819), to provide a significant improvement to the Newcomen design and to launch upon the engineering world a range of new concepts that would underpin the expansion of mechanical engineering through science and technology. Watt proposed the concept of ***horsepower***, the ***SI unit of power*** and the ***watt*** as a unit of power. In all of these things he was to mechanical engineering what Newton had been to gravity, the laws of motion, and the inventive process.

Born in 1736, James Watt began by making instruments but, lacking the appropriate opportunities, he could not progress his craft until he was offered an opportunity to set up a small workshop at the Macfarlane Observatory in the University of Glasgow. From there, he became friends with the noted chemist and physicist Joseph Black (1728–99) and the remarkable economist, philosopher, and author Adam Smith (1723–90). Watt quickly became an established metalworker and instrument maker. Inventive by nature, the young Watt was encouraged by his friend the physicist John Robison (1739–1805) to look at ways in which steam as a means of power could be developed in a machine superior to that of the established Newcomen engine. In fact, in 1763, Watt was asked to repair a model of the Newcomen engine that had been brought to the university, and from this detailed examination of its working principle, he came up with an improvement.

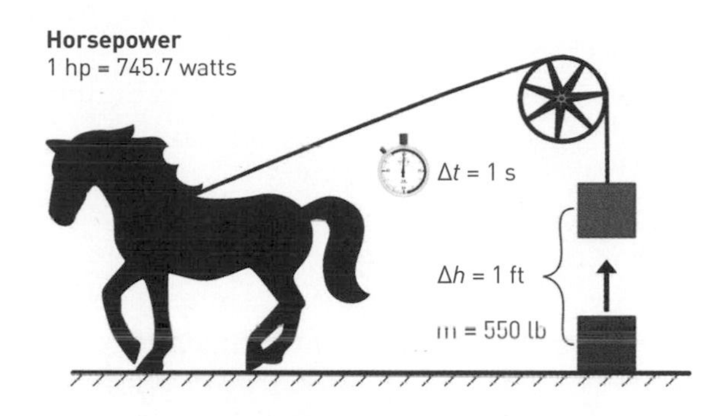

Bequeathed to future generations, Watt's imperial horsepower calculation by which the power of a steam engine, and later the reciprocating engine, was measured.

Watt decided that the operating principle of the Newcomen engine was limited by the need to repeatedly heat and cool the cylinder, a procedure that wasted most of the ***thermal energy*** rather than converting it into ***mechanical energy***. In fact, Watt calculated that approximately 75 per cent of the thermal energy was consumed by heating the cylinder on each cycle. Within two years, he had worked out that what the engine required was a separate cylinder in which the steam could condense away from the piston, thereby maintaining the temperature of the cylinder at the same temperature as the injected steam by means of a steam jacket. In this way, only a little energy would be wasted by absorption in the cylinder on each cycle.

James Watt with a model of his steam engine by the artist James Eckford Lauder.

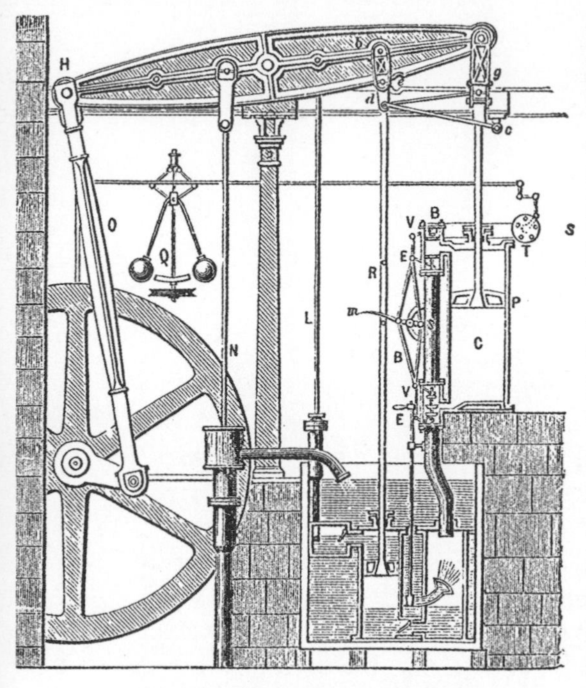

A schematic of a steam engine designed by Boulton & Watt, England, in 1784. Annotation: B steam valves (input); C steam cylinder; E exhaust steam valves; H connecting rod link to beam; N cold water pump; O connecting rod; P piston; Q regulator/governor; R rod of the air-pump; T steam input flap (controlled by governor (Q); g link connecting piston (P) and beam via parallel motion g-d-c; m steam inflow lever worked by the air-pump rod (R). From a diagram published in A History of the Growth of the Steam Engine *by Robert H. Thurston (1839–1903).*

Watt's legacy

What was required turned machining on its head and made skilled craftsmen out of metalworkers who doubled as boilermakers—skills hard enough to learn but of a very different level of precision to that required for the finely worked parts of the Watt steam engine and its separate condenser. In developing this technology, Watt virtually wrote the book on thermodynamics, although he was arguably more a practical man than a scientist. The story of how his device represented a turning point in industrial opportunities is long and cannot be covered in detail here. It is, however, important to appreciate how much of game-changer his invention was, touching so many new techniques and requiring scales of perfection, accuracy, and quality that were unnecessary with early Newcomen and Smeaton atmospheric engines. Here at last was an efficient steam engine—a design that could totally and utterly transform the way humans worked and what they could build.

James Watt also gave a lexicon of terms and measuring standards to the new world of the late 18th century, in time for a wide range of novel and unique applications. For example, he coined the term "horsepower" to measure the output of steam engines by comparing them to the power of draft horses. This would come down through the next 200 years and more, converted into metric from imperial and also expressed in SI units. By the early 21st century, the watt was almost universally adopted worldwide as the SI measurement of power.

DEFINING THERMODYNAMICS

We have already seen that the fundamentals of thermodynamics which underpinned all these developments were mapped out when Guericke built a vacuum pump in 1650. Everything that followed started the process that is still underway today, defining physical laws that have extended far beyond the basic principles laid down at various institutes during the 19th and 20th centuries. With the work of Savery and Newcomen, the basic principles of thermodynamics were demonstrated, but their machines were crude and today would be considered inefficient. Binding the concept of heat capacity and latent heat during the mid-18th century, Professor Joseph Black at Glasgow University developed thermodynamics while working with James Watt, but as we have seen, it was the latter who contributed the external condenser and, while transforming the practical possibilities, provided the stimulus for a broader and more scientific method for defining the three laws that we recognize today.

> **Calculating horsepower**
> Using the imperial measurements of the 18th century, Watt calculated that a pony could lift 220 lbf (0.98 kN) a distance of 98.4 ft (30 m) per minute over a four-hour working period. He scaled this up by 50 per cent to arrive at the conclusion that 1 horsepower = 33,000 ft lbf/min.

The first law of thermodynamics

The *first law of thermodynamics* expresses a derivative of the **conservation of energy**, which is in itself a fundamental principle relevant to a wide range of scientific disciplines. It states that the total energy of an isolated system remains the same (hence, it is "conserved") and that energy can neither be created nor be destroyed. Not to digress too far, this "law" prohibited completion of mathematical formulae expressed by astronomers supporting the "continuous creation" of matter in the universe, an idea prevalent in the 1950s; Einstein showed that matter and energy are interchangeable and this did not permit continuous creation.

The get-out clause for engineers stated by the first law of thermodynamics, however, is that while the total energy in a system cannot be created or destroyed, it can be transformed from one form to another.

First law formula
The first law of thermodynamics is usually formulated as $\Delta_{Usystem} = Q - W$, where $A_{Usystem}$ is the change in the internal energy of a closed system, Q is the heat energy supplied to the system and W is the thermodynamic work done by the system on its surroundings, which carries a negative sign.

As well as the conservation of energy, the first law also embraces the concept of internal energy, which in its relation to temperature has three separate but equally observed elements:

1. Kinetic energy, if the system as a whole is in motion
2. Potential energy, if the system is existing within an imposed field, such as gravity
3. Unique internal energy, which is a fundamental part of thermodynamics

This defines the specific aspects of the first law and separates it from the more general law of the conservation of energy. Work itself is defined as a process of transferring energy in ways prescribed by mechanical forces, and these can range from a stirring mechanism driven by a driveshaft to a piston used to compress a fluid. In this way, added work can increase the energy of a system, and because that is a means by which energy transfer can take place, work can derive from kinetic energy, potential energy, or internal energy. Note that the first law also embraces heat flow as a form of energy transfer.

The second law of thermodynamics

The *second law of thermodynamics* has special significance for mechanical engineering. It posits the irreversibility of natural processes and implies ***entropy*** within the thermodynamic equilibrium. In these expressions, entropy as defined here establishes an equitability between microscopic and macroscopic quantities, that number defined by Ω. But the essence of the second law states that if two separate and isolated systems, each in thermodynamic equilibrium with itself, are made to interact, they will reach a mutual equilibrium. This implies that the entropic state of each system when summed is no greater than the total entropic value of the combined systems. It can be less, but it cannot be greater.

In direct application, the second law qualifies what engineers had known all along: that when two systems of different temperatures come into contact, heat will always flow to the colder system. In the broader application, for defining unknown parameters of the microscopic, extrapolation from a known Ω of the macroscopic allows a determination of that microscopic value.

The third law of thermodynamics

The ***third law of thermodynamics*** is a statistical law connected to entropy and states that it is impossible to reach a temperature of absolute zero (0 K or Rankine/ -273.15°C) where there is no activity. It presents a reference point for this and, while generally considering it to be the definition of entropy, it has been moved around, entropy itself sometimes standing as the "third law" and sometimes amalgamated with the second law.

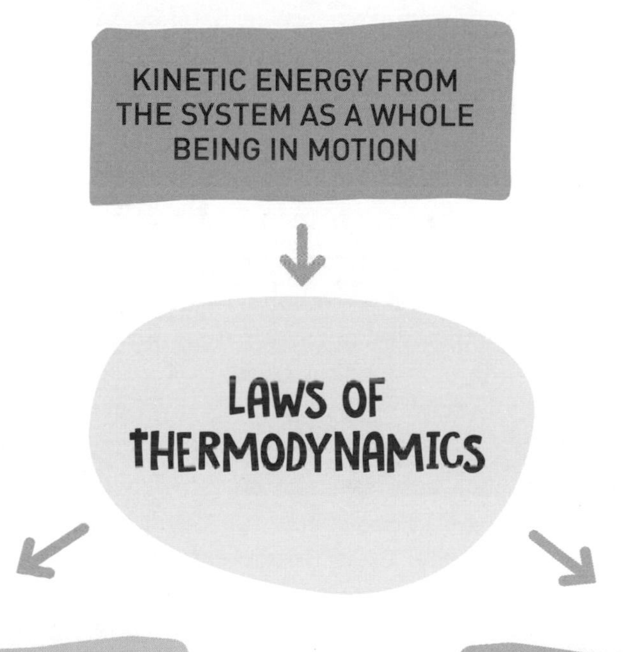

Cumulative knowledge

In tracking their definitions, we learn that the laws of thermodynamics were not written down by one person at a specific time on flashes of inspiration or by trial and error, but rather through the combined and evolving theories of various bodies and institutes over a period of around 150 years, and also that considerable debate surrounds the definition of these "laws."

As defined here, the first thermodynamic textbook was written by William Rankine (1820–72) in 1859, bridging the disciplines of physics and engineering and opening the door to a completely new field of mechanical engineering, shaping practices and applications that had already been in development for 150 years. But it was left to James Clerk Maxwell (1831–79), more famous for unifying the forces of electricity and magnetism, Ludwig Boltzmann (1844–1906) and Max Planck (1858–1947) to lay down the foundations of ***statistical thermodynamics*** and to bequeath a new way of thinking about the unification of mathematics and engineering involving heat engines and power.

Powered by steam but reserving sail as a precaution against failure, the White Star Line's RMS Oceanic *boasted four-cylinder compound steam engines, a type in which the engine has more than one stage for recovering energy from a working fluid.*

ACCELERATED PROGRESS

The pace of progress after the emergence of the Watt engine was phenomenal. When Watt had set out from Greenock at the age of 26 to travel to London, he went on horseback, as was common practice. He traveled via Coldstream in the Borders, Newcastle, and thence to the Great North Road—a dusty unpaved track, wide enough for carriages and coaches. It took the young James 12 days to get to London, and his luggage went by sea because the roads were too unreliable. What were then dirt tracks, wagonways, unnavigable lanes, and flooded ditches would be transformed within a generation through the invention of the steam engine. And when steam power was put to wheeled wagons previously pulled by ponies along tracks, the age of the steam locomotive had arrived.

All the methods, calculations, machining of finely crafted parts and trials of countless inventions for road, rail, and sea would change the world forever. Through these advantages and because of the contribution from engineers and inventors, Britain would acquire an empire larger than any country had seen before, and a new breed of scientists, technicians, designers, and businessmen would feed a new generation of mechanical engineers. From the essential principles laid down by inventors, thinking ahead of their time, would flow a new Europe and a new range of capabilities that would eventually usher in the age of the internal combustion engine at the end of the 19th century. With that would come the motor car and the aeroplane and, using the same principles of thermodynamics, the jet engine and the rocket motor. And from those would come passenger-carrying supersonic airliners and in the same decade that those began to fly, the first humans to land on the Moon in July 1969. Thermodynamics defined all these inventions.

After providing the engine for a steam-powered army tank in 1918, five years later the Doble brothers produced a steam-powered car, one example of which is preserved at the Henry Ford Museum in Detroit, Michigan.

Chapter Four

MECHANISMS

Predicting and Mitigating Failure—Gears and Cogs—Cams and Followers—Other Forms of Gearing

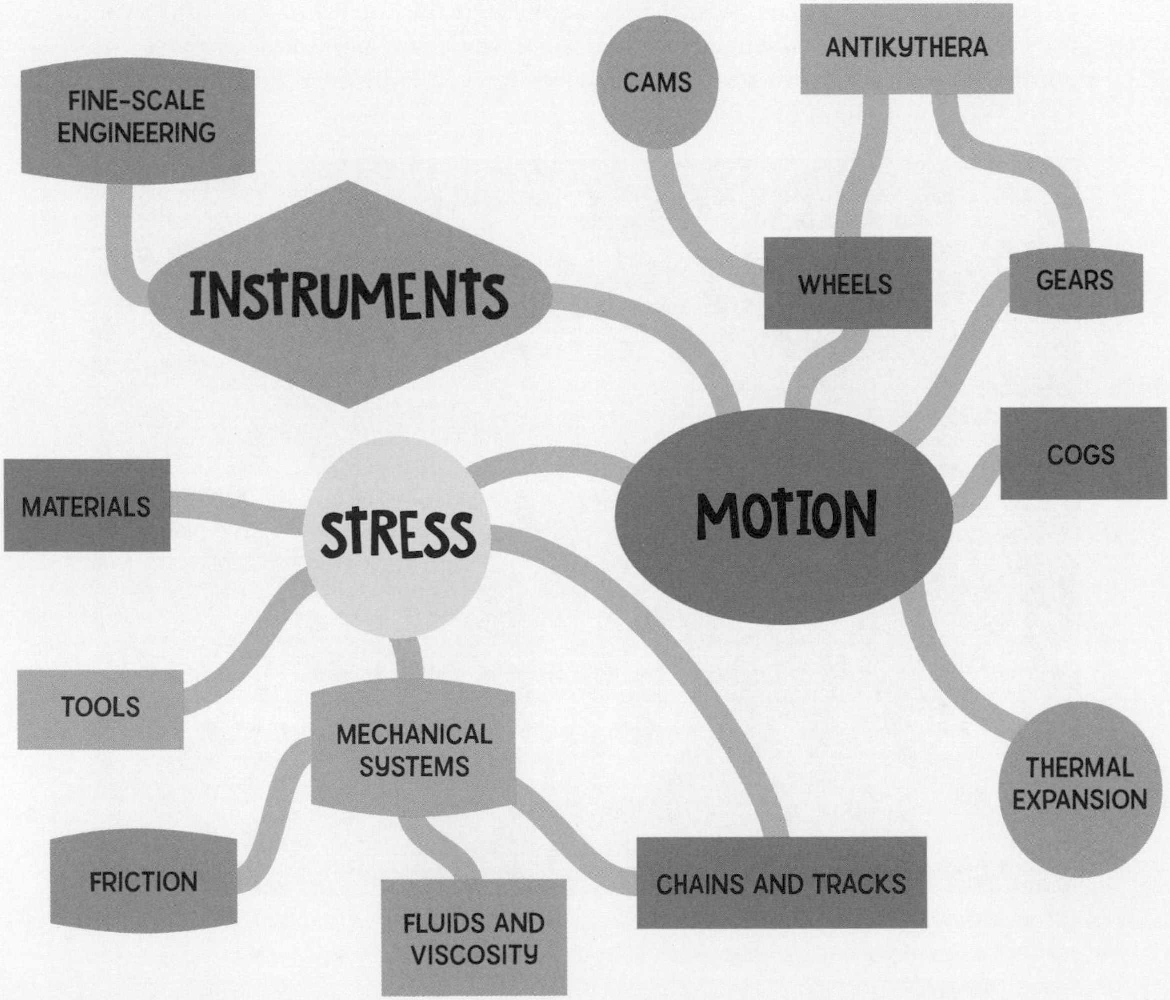

CAMS

ANTIKYTHERA

FINE-SCALE ENGINEERING

INSTRUMENTS

WHEELS

GEARS

COGS

MATERIALS

STRESS

MOTION

TOOLS

MECHANICAL SYSTEMS

THERMAL EXPANSION

FRICTION

FLUIDS AND VISCOSITY

CHAINS AND TRACKS

PREDICTING AND MITIGATING FAILURE

So far, we have looked at forces and the use of empirical thinking as it was replaced by the scientific age of learning, laws, and calculation. The capacity of the human brain for abstract thinking is like a machine itself, from which has developed a base of knowledge in the engineer for taking natural materials and fabricating tools into simple machines, and from those on to complex, compound systems that operate integrally and as a unit. But we have not yet examined the challenges posed by taking relatively simple devices such as wheels, levers, and linkages into compound systems with interconnecting mechanisms, capturing energy and liberating it for productive work.

Key to understanding limitations on what can be done to add value or leverage to simple tools and basic machines is measurement of ***stress***, which, in mechanical engineering, defines the internal forces that adjacent particles in a continuous material exert on each other. Coupled to stress is strain, the measure of the deformation in the material caused by stress, which in mathematical terms is represented by the Greek sigma sign (σ). Understanding stress (by calculation) through knowing ***strain*** (by measurement) enhances possibilities for increasing the work potential while minimizing failure, or its propagation, and understanding limits imposed by the structural nature of materials.

In practice, Newtonian physics comes into play, dictating, for instance, the individual particles of a member or pressure vessel under stress as particles press not only upon each other but also upon the restrictive envelope of, say, a beam or a containing vessel for a fluid—be it gaseous or liquid. Thus can Newtonian law help determine failure points or probabilities through probabilistic calculation derived from analysis. Stress can be caused by external influences such as gravity or acceleration and deceleration, or by contact, pressure, or friction. Deformation of a solid object itself is likely to cause strain due to an ***internal*** (***elastic***) ***stress***, which can be thought of much as the reaction in a spring. But it can also be caused by gradual and deformational pressure known as viscous stress. It is these two effects—elastic and ***viscous stress***—that are known as ***mechanical stress***.

There are, however, subtle variations, including ***built-in stress***, pre-stressed concrete or tempered glass for example, or stress caused by subtle changes in environmental conditions, and these include changes in temperature or chemical composition.

In mechanical engineering, particularly where heat engines are involved, ***compound stress*** can be a product of all of the above! In which case, stress has to be calculated and the effects measured for machines called upon to change their state of work to perform the job they are designed for.

Quantification of stress is vital for an effective design. However, not until the 17th century could it be defined, raising it out of the empirical age. For engineering purposes, stress is expressed through the traction vector T, which is defined as a force F between adjacent parts across an imaginary surface S, the sum divided by the area of S. Stress in a fluid, which is at rest in an Earth gravity environment, is perpendicular to the surface. This is expressed as pressure. With a solid structure, as well as in the flow of a fluid, F is not necessarily perpendicular to S, in which case it is regarded as a vector quality and not scalar.

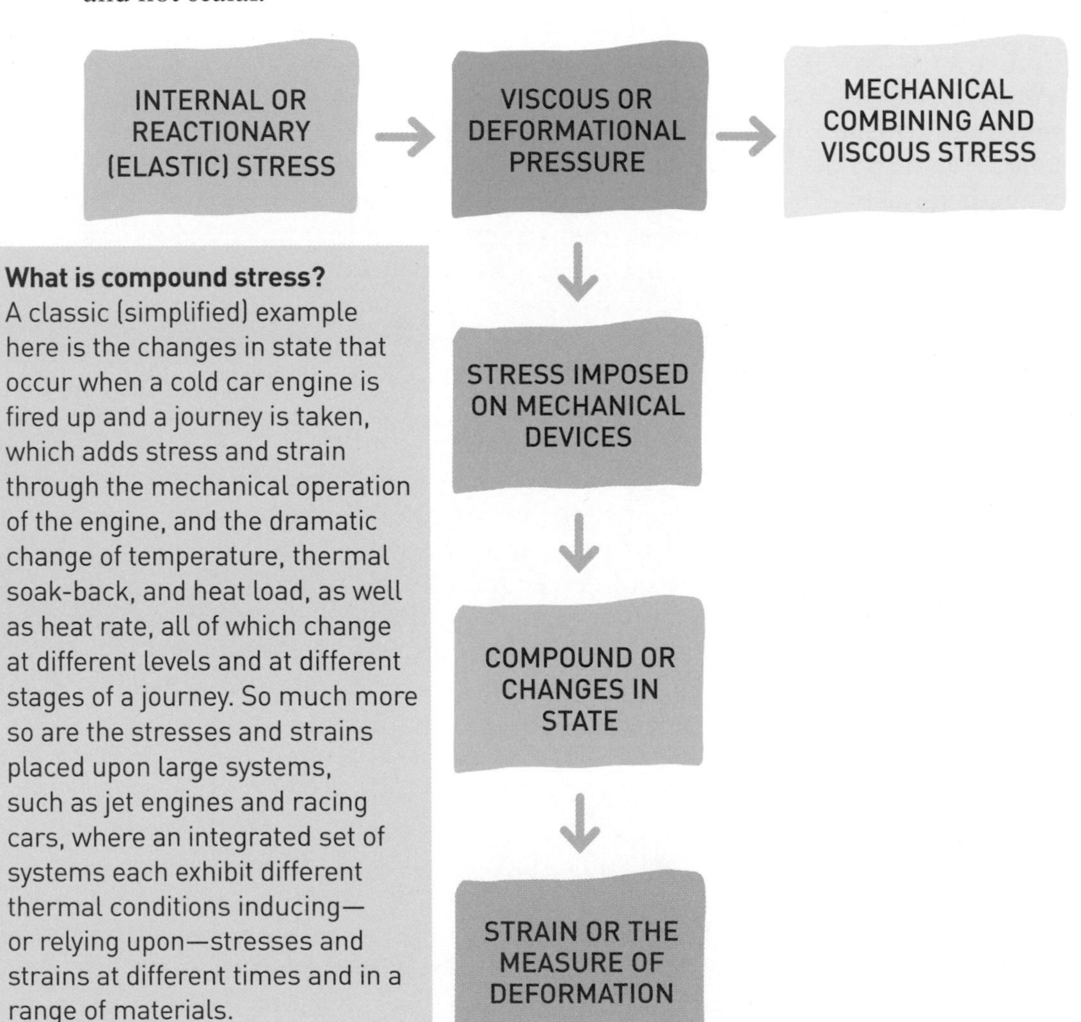

What is compound stress?
A classic (simplified) example here is the changes in state that occur when a cold car engine is fired up and a journey is taken, which adds stress and strain through the mechanical operation of the engine, and the dramatic change of temperature, thermal soak-back, and heat load, as well as heat rate, all of which change at different levels and at different stages of a journey. So much more so are the stresses and strains placed upon large systems, such as jet engines and racing cars, where an integrated set of systems each exhibit different thermal conditions inducing—or relying upon—stresses and strains at different times and in a range of materials.

A stress test on material is described by a tensor and this is a linear function relating the normal vector n of a surface *S* to the stress *T* across *S*. From this, the variation in stress across or within a body is regarded as a ***"time-varying" tensor field*** that can change with time and across different parts of the body. In that regard, an integrated tensor field maps strain levels as reflected in the overall stress.

GEARS AND COGS

Every time we use a wheel, a wedge, a ramp, or a hammer (to name but a few examples), we impose stress that is not usually limiting in what we do with the tool. But when we combine these—each with their own forces subservient to fields and physical laws—we impose ***specific stress***, which is not that for which

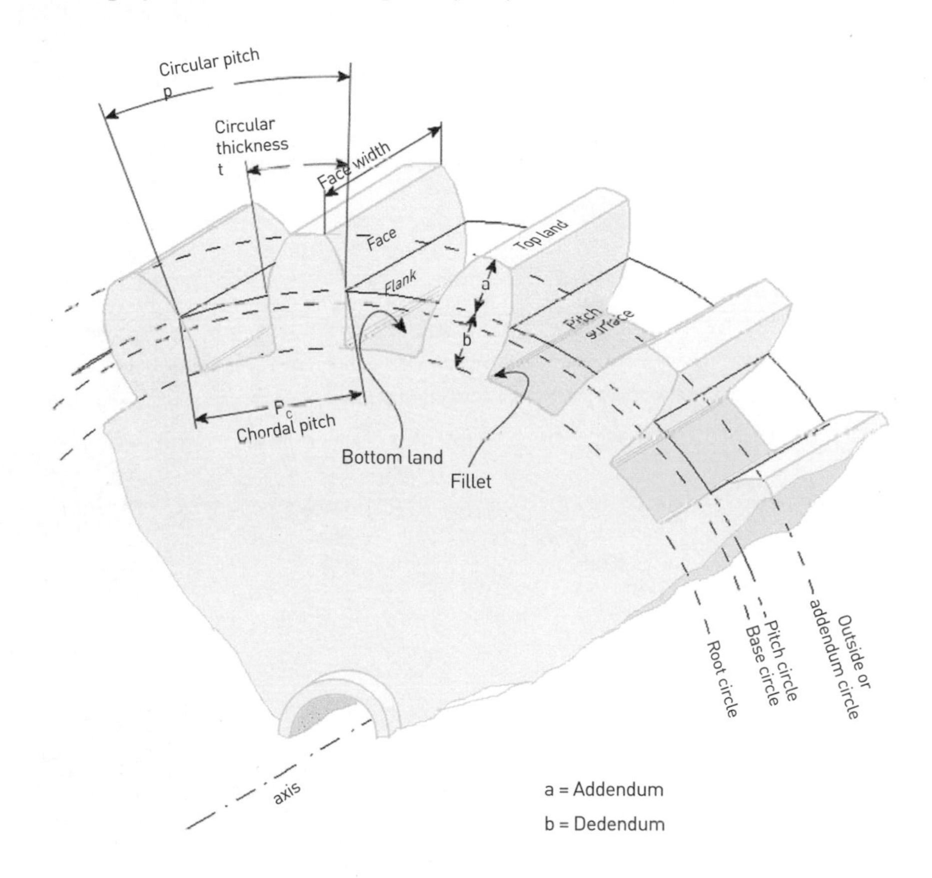

Names and phrases for identifying various parts of a cog and tooth-gear arrangement universally applied throughout the world of mechanical engineering.

the tool is designed. Such is the case with the ***wheel***, which we will examine both here, in relation to ***cogs*** and ***gears***, and in the next chapter, where we will look at several examples of how machines are kept moving through linkages, cams, conduits, pipes, and clutches. However, first we need to look at wheels, gears, and cogs in greater detail.

It is in examining the real application of energy and power to a gear that we realize the advantage that it provides. But, like a length of string, there is no strict definition of what a gear is. In fact, the very term "gear" implies a ratio of different outputs, although strictly speaking that is not a gear but rather an inclusive term for a set of separate gears, each of a different size, to work as a self-contained machine to control output according to the way the gears work together. A gear is not a ratio of output, or a comparative term as to "which gear to pedal." It is a rotating machine with an independent number of cut teeth that creates mechanical advantage through producing a change in torque—which is something we have encountered before.

A replica of the ancient Greek Antikythera, a highly sophisticated mechanical calculator using cogs and wheels, believed to predict the motion of the Sun, the Moon, and known planets as interpreted here depicting the cyclical and epicyclic motion of the cogs and wheels.

In almost the simplest of machines that it is possible to envisage, a gear has cogs designed to mesh in with another cogged wheel to transmit torque. This is almost as old as empirical design itself. In fact, the earliest known gearing system was used in China more than 2,350 years ago, while the earliest application in the West is in the Greek Antikythera mechanism of probably contemporary date—a device used for making calculations of the movement of major bodies in the night sky and possibly for predicting eclipses and other astronomical phenomena. And again, we note that Hero of Alexandria records the use of gears in AD 50, as he acknowledges the intricate geared devices of Archimedes during the 3rd century BC.

Gears were clearly a functional key to many small instruments, including the first geared mechanical clocks, which appeared in China around the 8th century. However, their use in devices for lifting water is again attributable to the ubiquitous al-Jazari of the 13th century. A few hundred years later, in the 16th century, a ***worm*** gear device is known to have been adopted for a roller cotton gin.

Compound materials used in a medieval application of cogs and gear drives on a cast-iron mortise wheel with wooden cogs powered by a water-driven cast-iron gear wheel.

Worm gears and their functions

A worm gear is a circular screw that is placed so that the grooves mesh with the teeth of a cogged wheel, converting motion through an angle of 90 degrees. In fact, there are three types of worm gear:

1. ***Non-throated worm gears***, which have no groove.
2. ***Single-throated worm gears***, with only the worm gear having a groove.
3. ***Double-throated worm gears***, which have both gears throated and therefore accommodate maximum loading.

A unique advantage with the worm gear is that it is non-reversible—a characteristic that has added mechanical as well as safety aspects.

The ability of a worm gear to compensate for asymmetric alignment creates opportunities in engineering design, and it has proven itself in several applications, including in sailing ships to more effectively control the rudder, and in milling heads. It also found value in the design of early motor vehicles, which, when they suffered a blown tyre, tended to pull the steering mechanism toward that side. By applying a worm screw, the effect was neutralized. The value of this capability led to refinements, including recirculating ball bearings that limited frictional forces and reduced wear, increasing the steering effectiveness and reducing the risk of failure by reducing stress.

Like many innovative devices in mechanical engineering, a basic operating principle can find wide application in a variety of technologies. Because it is a compact means of decreasing speed while increasing torque, the ***worm drive*** is frequently applied to electric motors, which inherently in operation have high speed and low torque. They are effective, too, when applied to motor vehicles, providing a better ground clearance for working trucks when placed high above the crown wheel of a differential (more on this later). By relocating the effective operation of the ***differential*** gearing, some cars in the 1960s had the worm gear below the differential so as to significantly lower the floor height.

In everyday application, stringed musical instruments use worm gears to adjust the tension in strings; in such application this is called a ***machine head***. Plastic worm gears are commonly used in very small, battery-operated devices—particularly toys—so that an output can be obtained with a low angular velocity, producing a "geared" spin rate that's less than the motor itself.

The advantages of coupling gears

Clearly, in this description we have been blending gears with ***gear-train devices***, and indeed the coupling of separate gears into a working subset of a larger machine is the most advantageous use of the basic gear and cog. In addition, a ***mechanical advantage*** is obtained. When two gears are to be aligned, the number of teeth is proportional to the radius of the pitch circle ensuring that the combined pitch cycles integrate with each other and without slippage. The advantageous work output is determined by the speed ratio of a pair of meshing gears, and this can be computed from the pitch circles and the ratio of the number of teeth on each gear. It is the respective number of teeth on each circle that determines the geared speed of the output wheel and the ratio of the numbers of teeth on each wheel.

Calculating the velocity of a geared wheel

The velocity v of the point of contact on each pitch circle is the same on both gears and is shown by $v = r_A \varpi_A = r_B \varpi_B$, where input gear A with radius r_A and angular velocity ϖ_A meshes with gear B with radius r_D and angular velocity ϖ_B.

Bicycle gears

The sprocket itself has been developed into a small working machine. Its ubiquitous use as a bicycle chain is simple enough but here it has been possible to modify the gear ratio by varying the diameter of the sprockets—adjusting up or down the number of teeth on either side of the chain. The most developed form is that of the derailleur system, which consists of a chain, multiple sprockets, and a small mechanical tool for remotely transferring the chain up or down the gear ratios while the chain itself is in continuous motion. Gear ratios on the bicycle are moved up or down by shifting the chain to left or right, and this is usually controlled via a Bowden cable operated by the cyclist. A Bowden cable consists of an inner cable moving relative to a fixed outer tube to actuate a mechanical process.

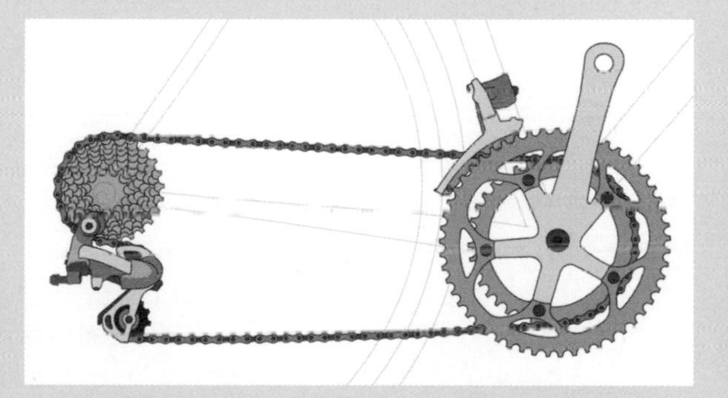

One common application of the gear drive chain with cog changes possible during wheel rotation, the derailleur bicycle drive train is a popular mechanism for changing to a higher gear during wheel rotation by a lever that shifts the chain to an adjacent cog.

Helical gears have been applied to many different systems, here displaying a parallel (top) and crossed gearing.

CAMS AND FOLLOWERS

There are various ways in which work can be transferred through a mechanical system, and as well as gears and gear trains these include ***cams*** and ***followers***. A cam is technically defined as a device that imparts a prescribed reciprocating motion to a follower, the latter being the output element of a cam mechanism, which is in contact with the cam profile. The motion is determined by the shape of the cam surface, known as the ***profile***, and the geometry of the contact surface with the follower.

Rotating cams are most commonly used in engines but they can also be of linear form (wedges) and cylindrical, in which the end of the follower runs in a groove profiled to give the desired motion. But there are other types, too: an ***eccentric cam*** is circular in form but rotating off-centre, while ***rotating cams*** are frequently operated by a ***camshaft drive***, which can take the form of a ***toothed belt*** or a ***sprocket chain***.

The most commonly encountered cam/follower system is that employed in ***reciprocating engines*** (in which pistons connected to a crankshaft move back and forth in a cylinder). Cams and followers allow complicated motion to be achieved in a manner that's difficult to readily accomplish with other systems. ***Internal combustion engines*** employ cam/follower combinations to operate the inlet and exhaust valves in a precise and co-ordinated manner. One dual-purpose cam involves a cylindrical cam with a helical groove, which, as it is tracked by a follower, changes a rotational force to a linear movement. The ability to have two adjacent wheels rotating at different speeds is necessary for some fixed machines, and it is here that gears, cogs, and followers come into play. But there is a use of gears that is driven by efficiency and safety: the car.

When two wheels of a cart are turning on an axis perpendicular to the direction of travel, both are moving at the same rate of rotation. When the cart is made to turn, the wheel on the inside of the turn is moving slower than the outer wheel because the latter has farther to travel. This works without stress if each wheel is independent of the other and not fixed to a connecting shaft, instead being free to rotate at the speed required to complete the turn. Early motor vehicles had the engine connected to only one of the rear wheels, leaving the other to freely turn at a greater or lesser rate. But that meant that the total work done by the engine had to be delivered through one wheel, which caused reduced performance on the road with considerably reduced traction. The solution was the ***differential***, which allowed power to be applied to both rear wheels without compromising the traction of each wheel as it turned at its own rate.

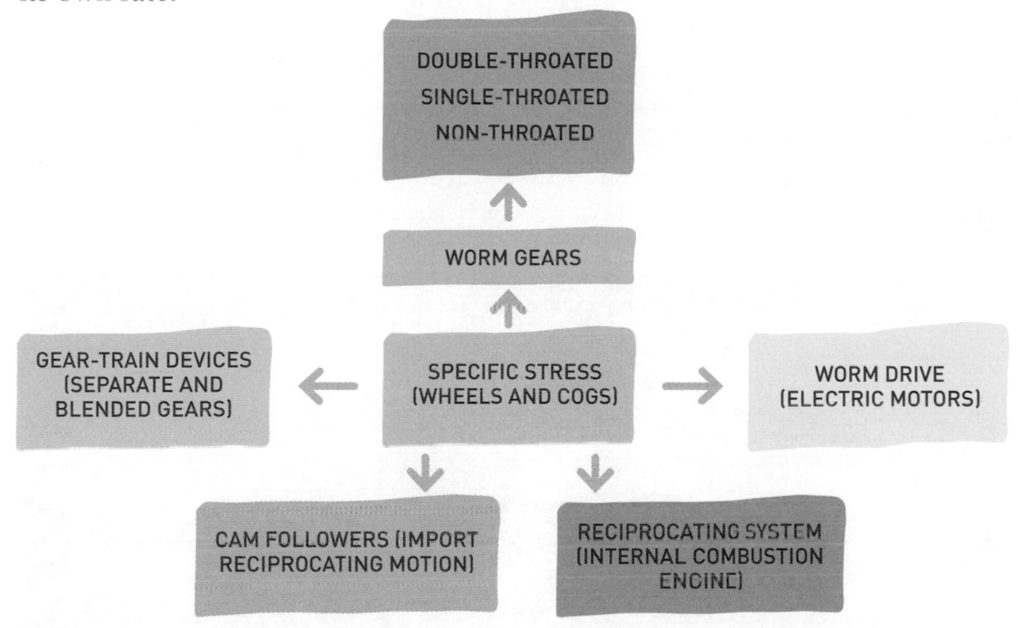

Doing things differentially

The problem with applying power to both rear wheels while allowing the wheel on the outer radius of a turning circle to move faster than that on the inner circle is that a single axle is no longer possible. To apply power to the rear wheels, the ***driveshaft***, which takes energy from the engine, is connected to each wheel shaft but with a ***transverse ring gear***, which also works as a reduction gearing. Essentially, it consists of an input (the driveshaft) and the two output shafts, to each of which is connected a wheel. For ***motive traction***, the driveshaft converts rotational motion along the longitudinal axis of the vehicle to rotational motion along the transverse axis—a change in rotation of 90 degrees. The ratio of the speed of the two drive wheels is defined by the ratio of the radii of the paths around which each wheel is rotating. The value of that is defined by the track

of the vehicle, or the distance between the two wheels, while the radius of operation in the differential will also depend (for a given radii of curve) on the wheelbase of the vehicle, the distance between the front and rear wheels.

The differential, converting energy from the driveshaft to the wheels on a Skoda 422.

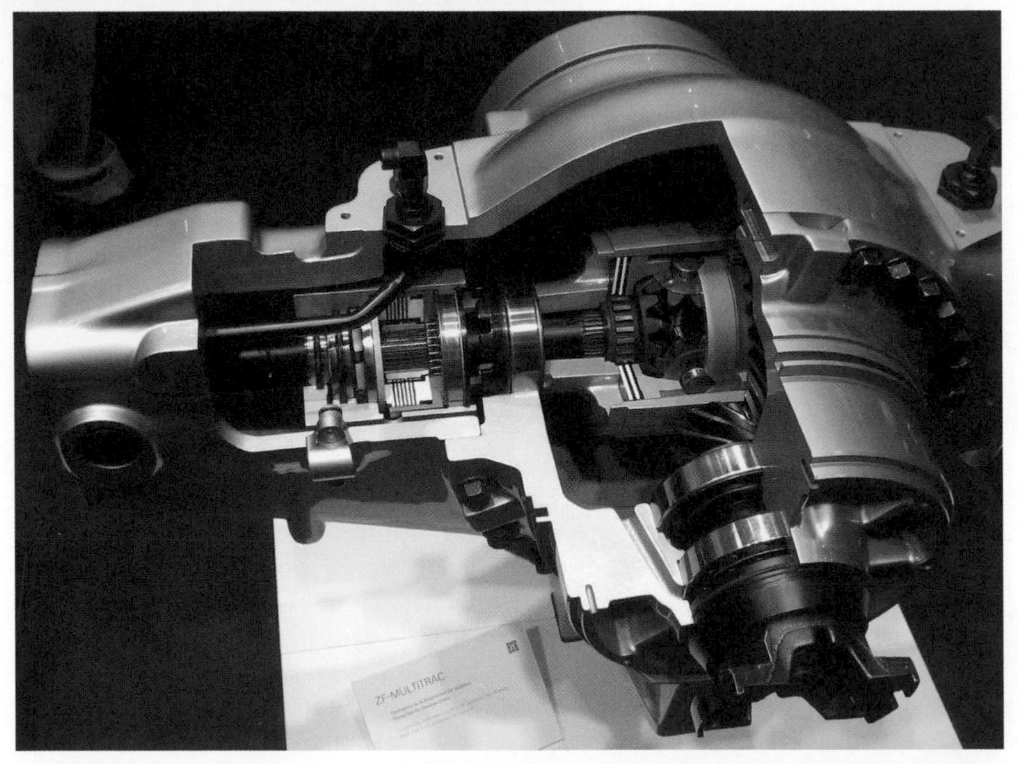

The ZF differential with the driveshaft entering at the front and converting the axis of rotation through 90 degrees.

OTHER FORMS OF GEARING

Of course, technology moves along and there have been innovations, including ***epicyclic/planetary gearing***, which effectively divides the torque between the rear and the front wheels. Using precisely the same theoretical principles as the conventional differential, this consists of two gears mounted so that the center of one gear revolves around the center of another, with a carrier rigidly connecting the centers of the two gears and rotating in unison to carry one gear around the other. Logically, the gear that revolves around the center is known as the ***sun gear***, while those rotating around the inside of the outer wheel are known as ***planetary gears***.

There are many forms of cyclic and epicyclic arrangements, the most simple in everyday use being the hand-spun salad shaker, where the planetary gear moves through an offset alignment with the rotating handle to invoke inverse centripetal force to shed water away from the moist salad leaves.

In effect, there is no limit on the number of planetary gears in an epicyclic system, and a ***compound planetary system*** is one where there are more than two planets in a mesh, or within each other in a mesh train. Compound gears can also be applied in a ***stepped-planet configuration*** where several epicyclic configurations are contained together in a vertical stack, arrangements where the step effect is the same but in much less space, in some cases occupying

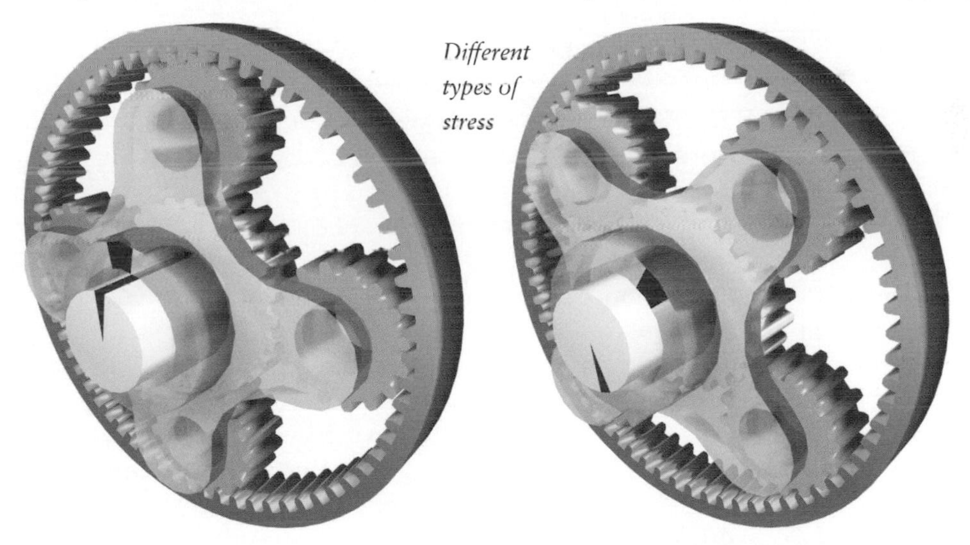

Epicyclic gearing provides torque asymmetrically, as shown here where the input shaft is the green hollow tube, the low torque being the yellow one. The high output torque is pink. The force applied in the yellow and pink gears is the same, but the torque on the pink arm will be two or three times as high.

Funicular systems
Very steep rack rails, with added safety precautions, are found in exceptionally steep rail systems, such as the funicular, the steepest of which, at a gradient of 47 degrees, is located in Stoosbahn, Switzerland. But a funicular is unique in that as one coach descends it pulls the other up, where only the contact with the rail involves a rack and pinion device.

half the diameter, therefore opening up a wider range of applications, even down to instrument level.

At the other end of the scale is the ***rack and pinion***, essentially a class of gears or linear actuator that involves a circular gear known as the pinion engaging with a linear gear known as the rack. All it does is translate rotary motion into linear motion by rotating the pinion wheel into the fixed rack, as utilized in so many examples of rack railway systems around the world. A conventional railway cannot operate at gradients of more than 10 per cent before the wheels start to slip, which is where the rack rail system comes into play, with a ratchet designed to prevent it slipping back down a steep incline.

In mechanical systems, there are several uniquely divergent applications of the same concept using precisely the same forces and gear/cog combinations. One is the modern battle tank, which evolved out of agricultural machinery using a track-laying system whereby the footprint of the track remains static on the surface as the vehicle rolls over it by means of sprocketed cog wheels. Essentially, the pinion uses rotary motion to rotate a gear wheel, which engages with a track wrapped around the drive system on a continuous loop. In effect, it reverses the functional work source of a conveyor belt, by allowing the machine itself to roll over a continuous belt rather than having a static machine carry people or objects along a belt.

In most applications, the continuous track comprises a solid chain track assembled from individual steel plates, but on modern applications the track can consist of synthetic rubber so as to inflict less damage to metalled road surfaces. Its application goes back into the mid-19th century and before, when agricultural vehicles adopted this system in order to reduce the pressure induced by wheels and laid a flat-footed tread rather than a rolling tread. Because of that, tracked vehicles are ideal in marshy ground or across snow-covered surfaces, and are widely applied to a range of civil and military vehicles alike. Subtleties of use provide different tread materials. Tanks traveling along paved roads will have had their steel treads capped (or replaced) by dense rubber treads; steel tread faces can sometimes carry lips on their leading edge to grip unpaved surfaces, but if operated on metalled roads would effectively tear up the surface.

FRICTION

Before we leave the basics of connected surfaces engaging work from energy and move along to some rather more sophisticated ways in which machines are kept moving, there is one more force to consider, which is vital in mechanical engineering: friction. Like everything else, the Greeks were fascinated by the different aspects of ***friction***, noting that less energy is needed to keep a moving object in motion than there is in imparting motion from a relative static position. Basic laws were written down by Leonardo da Vinci (1452–1519) in 1493, from where the science of tribology emerged, describing friction, lubrication, and wear. (Tribology is concerned with the science of interacting surfaces and is an essential part of engineering design across a wide range of specialties—be it static or in motion.)

There are two types of friction: ***dry*** and ***wet***. For our purposes here, dry friction is divided into ***static friction*** and ***kinetic friction***.

Friction formula

Dry friction is calculated by the formula $Ff \leq \mu Fn$, where Ff is the force of friction exerted by each surface on the other, μ is the coefficient of friction (the empirical property of the contacting materials) and Fn is the normal force exerted by each surface on the other (perpendicular to the surface).

In the formula, the normal force is the net force compressing the two parallel surfaces together, and its direction is always perpendicular to the surfaces. Under normal at-rest situations, the only part of the normal force is that due to gravity, so that N = mg, the magnitude of the friction force being the product of the mass of the object. In this case, the coefficient of friction of a heavy block will be no greater than the coefficient of friction of a small block, but the magnitude of the friction force is determined by the normal force and therefore by the mass of the block.

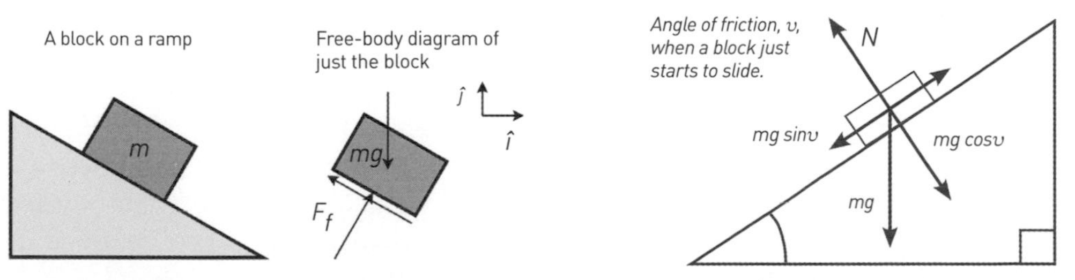

The free-body diagram for a block on a ramp with vectors shown by arrows indicating both direction and magnitude of forces, where N is the normal force, gravity is shown by mg, and the force of friction by Fr.

Symbolized by the Greek letter μ, the ***coefficient of friction*** (COF) describes the ratio between two bodies and defines the forces pressing them into contact. It is dependent on the materials measured: steel on ice has a low COF, while a rubber tyre on a paved road has a high COF. Measure scales grade the COF in a range from zero to more than one and are greater between surfaces of similar materials than between surfaces of dissimilar pairs. To move, the static friction must be overcome by an applied force. The maximum possible friction force is the product of the coefficient of static friction and the normal force, as in $F_{max} = \mu sFn$. But where no sliding occurs, the friction force can have a value from zero to Fmax.

Two objects moving together and rubbing against each other experience ***kinetic friction***—be they two steel wheels rubbing against each other or a molecule of air rubbing against a ceramic heat tile. The coefficient of kinetic friction is denoted by μk and is characteristically less than half the coefficient of static friction for the same materials. But some have suggested that dry metal surfaces show very little difference. Nevertheless, the friction force between two surfaces after sliding is the product of the coefficient of kinetic friction and the normal force, as in $F_k = \mu kFn$.

In several applications within mechanical engineering it is important to measure the ***angle of friction*** (in other sciences referred to as the angle of repose), which is the maximum angle at which one mass will begin sliding along the surface of another. The friction angle is defined as tan $\upsilon\omega = \mu s$, where υ is the angle from the horizontal and μs is the static coefficient of friction between the objects.

We have said little about ***fluid friction***—or ***viscosity***—but we will return to it and discuss its application later. Suffice to say here that viscosity is the friction that occurs between two separate and independent fluid layers moving relative to each other and in contact on their interfaces. This can relate to a fluid in any form, be it gaseous or liquid, as the dynamics are the same. But as with ***skin friction*** and ***lubricated friction***, these will come together when we examine the way we keep machines moving. In that, we will look at the role of lubricants and thermal control as it relates to active fluid flow and managed heat regulation—all the way from heating boilers to spaceships!

The way machine parts work together, move against each other, and enable practical work to be improved as a result has preceded the next phase of industrial production. The manipulation of molecular, even atomic structure to make materials and composites has transformed the scientific method in mechanical engineering.

Chapter Five

A CHOICE OF MATERIALS

Understanding Materials—Ceramics and Polymers—Glass—Composites—Material Failings

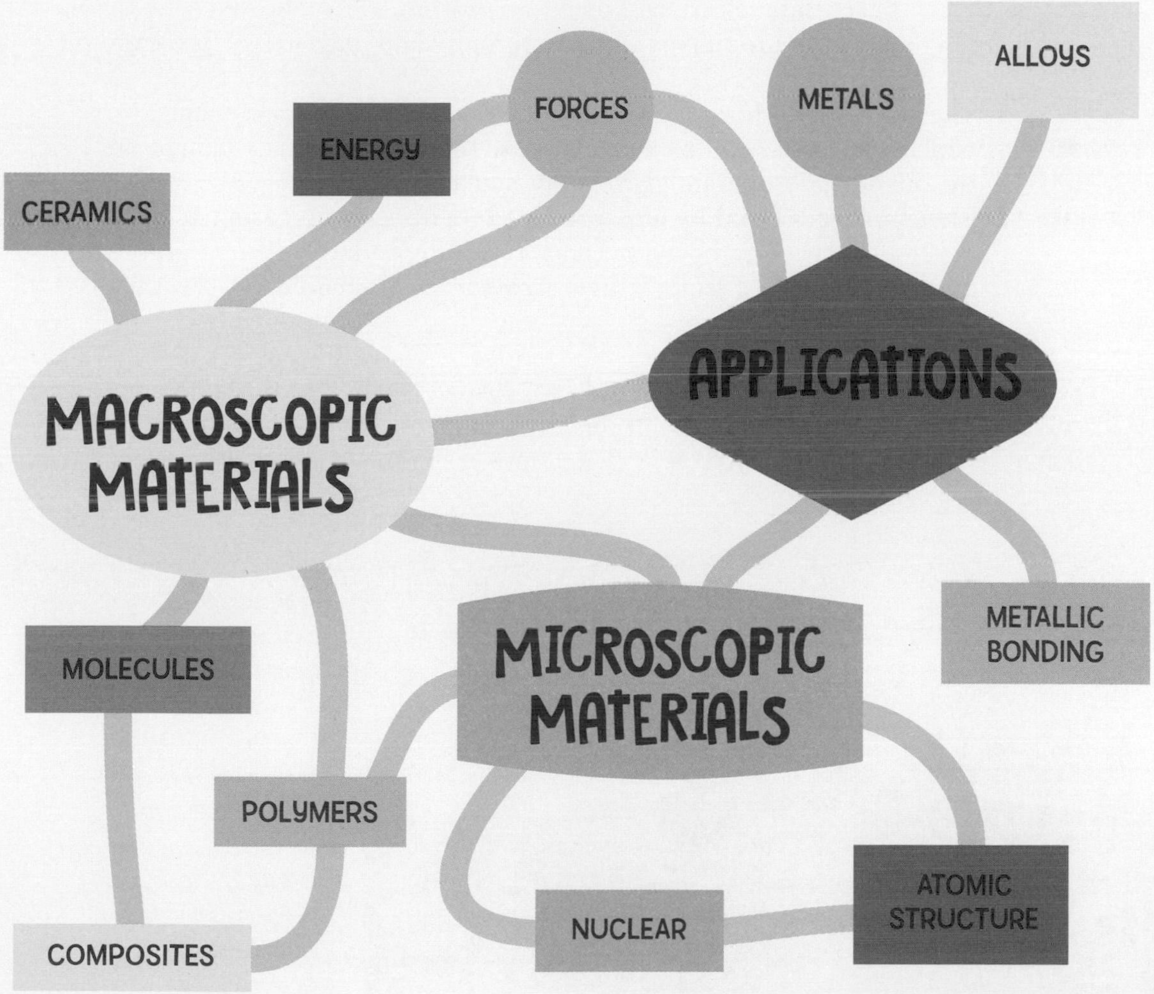

ALLOYS

FORCES

METALS

ENERGY

CERAMICS

APPLICATIONS

MACROSCOPIC MATERIALS

MICROSCOPIC MATERIALS

METALLIC BONDING

MOLECULES

POLYMERS

ATOMIC STRUCTURE

NUCLEAR

COMPOSITES

UNDERSTANDING MATERIALS

So far, we have been looking at solid mechanics—basic design analysis using forces, ***moments and couples***. We have examined how mechanical engineering began with simple tools and small machines, moving on to complex machines using scientific principles only within the last 300 years or so. But it has all been one-dimensional: basic principles of how components can come together to do work, how forces limit or open opportunities for even more complex machines, and how the manipulation of forces can create new technologies. However, in order to make machines move and operate, we also need to understand materials and their processing.

The properties of macroscopic materials are of great interest to engineers because only with the appropriate selection of them can machines be made that are capable of surviving in extreme environments—both internally and externally. Two examples:

1. The stresses and forces at work in an internal combustion engine, which are no less complex just because they are seemingly benign on the outside and operate in everyday machines.
2. The challenges faced by high-performance aircraft operating at supersonic speed or by a spacecraft returning to Earth through the atmosphere, where temperatures through friction can reach several thousand degrees.

Each requires the correct materials to survive high pressures, temperatures, and molecular reactions. Because of this, the engineer first needs to understand the materials they will work with.

Fabricated from materials designed to resist temperatures of up to 260°F (127°C), Concorde G-BOAB photographed at London Heathrow airport. This aircraft flew for 22,296 hours between its first flight in 1976 and its final flight in 2000.

Gaining a material understanding

Understanding materials starts at the ***atomic level***, which not only provides information about the ***basic structure*** of the material itself but also provides us with the ***attributes*** of the material and an understanding of how to control and modify it for specific purposes. All too frequently, materials need to be shaped into components with different processes for high-melt-temperature ceramics and low-melt-point polymers. Different processes would be used to make the metallic connectors on a flash drive, or to fashion a cylinder block for a large diesel engine. Because the processing of materials often changes their very nature at the atomic level, there is a need to understand how that will take place or, at best, predict through calculation what a specific action may produce.

We can assume here a basic understanding of the underpinning principles behind atomic structure and bonding. Engineers are interested in the way molecules are assembled from ***atoms*** and in the ***valency shells*** that determine how an element reacts chemically with another element—through the number of valency electrons of an atom. Fundamentally, the fewer electrons an atom has, the less stable it is and the more it is likely to react. The structure of the atom is itself best understood by the ***Bohr model***, but the presence of negatively charged electrons is a state of the atomic charge and not the certainty of a specific place on the outskirts of the nucleus; for as we now know, it is impossible to predict precisely where an electron is or where it is going, only that it presents the atom with properties that determine the element.

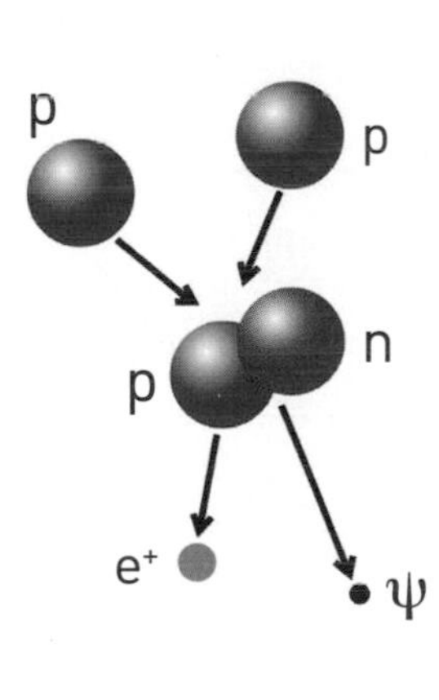

Utilized in a thermonuclear bomb, a nuclear fusion process forms a deuterium nucleus, consisting of a proton and a neutron, from two protons. A positron (e+), essentially an antimatter electron, is emitted along with an electron neutrino (y).

A model of the atomic nucleus, the fundamental building block of all materials, showing it as a compact bundle of protons (red) and neutrons (blue). An actual nucleus cannot be explained by a simplified diagram, but only by using quantum mechanics. In a nucleus that occupies a certain energy level, each nucleon can be said to occupy a range of locations.

Types of bonding

What is a bond?
The juxtaposition of similar or dissimilar materials, molecules, or atoms to strengthen or improve an existing structure.

Metallic bonding is achieved with a cloud of closely packed positively charged cations (positively charged ions) and the ***non-directional*** nature of the bonding allows metal atoms to pack more closely together. This results in high-density structures.

In ceramic materials, the bonding is ***ionic***, or ***covalent***, in which the electron is donated (or shared). As an example, a sodium chloride atom gives up a positively charged electron to become a positively charged sodium ion and in so doing the chlorine atom accepts the electron and becomes a negatively charged chlorine ion. As a result, the sodium atom evacuates its valence shell while the chlorine atom receives the single electron it needs to fill its own valence shell. Because of this transfer, the negatively and positively charged ions are attracted to each other and form a strong bond.

How polymers are formed
Polymers are formed from two types of bonding: covalent bonds between carbon atoms, which form the spine of polymers (monomers); and secondary bonding, much weaker, which is between adjacent polymer chains.

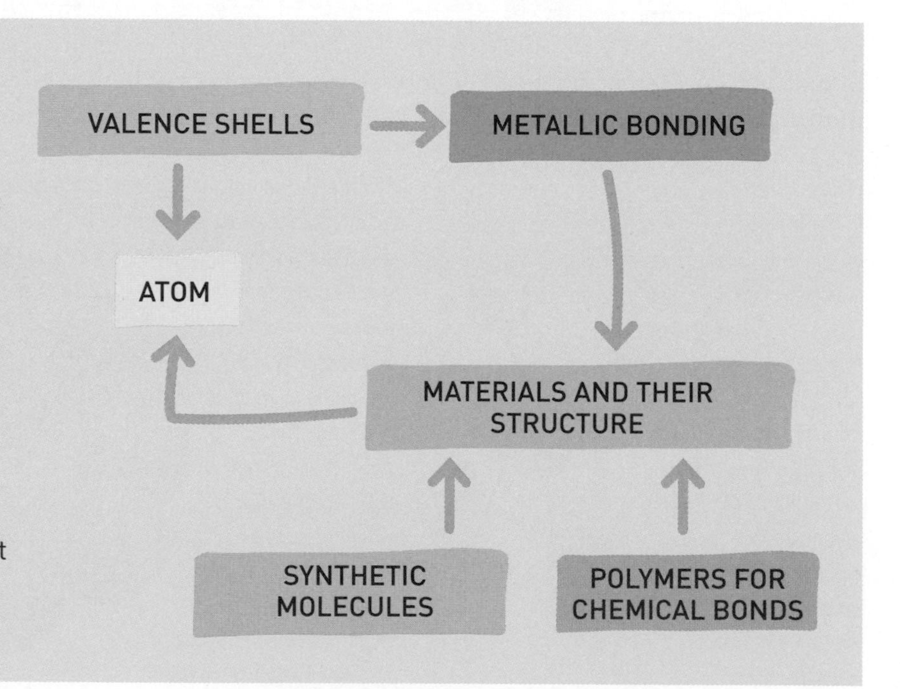

As *synthetic* or naturally occurring long-chain molecules, ***polymers*** are different again. They exist in a category that includes individual elements linked by covalent chemical bonds, including rubber, neoprene, nylon, PVC products, polystyrene, polyethylene, polypropylene, and silicone. They are the product of research in the chemical sciences beginning in the 19th century and are generally known as "***engineering plastics***," without any firm definition of the category. Regardless, they are a vital part of mechanical engineering in the scientific age.

But now, let's look at those bonds in more detail and see how they fit into the options for different materials available to mechanical engineers.

Metals

What is a metal?
A substance in its purest state of high electrical conductivity and malleability.

One of the oldest applications attributable to human ingenuity is the use of metals—in fact, metal was one of the earliest materials used to fashion tools and machines. Because metals have a non-directional bonding, which allows atoms to slide alongside and past each other, they have ***ductility*** that allows them to be deformed to a considerable extent without fracture. Moreover, because of delocalized free electrons moving easily under an applied ***electrical potential***, metals provide high electrical conductivity. Also, because metals have bonding that is strong and stiff, they offer high stiffness factors and high melting points. In addition, there are other aspects of metals that pertain to a wide range of applications.

Usually, in mechanical engineering, metals are found in a variety of ***alloys***.

Alloys

What is an alloy?
An alloy is a mixture of two or more elements where only the major component is a metal but where the combination has superior qualities for a given job than either can perform alone.

The definition of an alloy harks back to the fashioning of bronze from a mixture of copper and tin by humans' first smelters. On their own, most metals may be too soft, too brittle, insufficiently hard, or possess poor corrosion qualities. They can be too chemically reactive for use on their own and possess combinations of characteristics that are difficult to integrate into an engineering design, usually a machine. Each quality or characteristic can be inappropriate for a given application, and it is only through a selective menu of alloy options that the engineer can select appropriate materials based on metals of one or another type. Common alloys include ***steel alloys***, ***aluminium alloys***, ***copper alloys***, and ***nickel superalloys***.

Carbon steel alloys

Carbon steel alloys are a mix of iron and carbon and can have added elements, such as chromium, molybdenum, and nickel, which provide added strength and are generically known as "steel alloys." Stainless steel has about 11 per cent chromium, which provides a far superior resistance to corrosion, and this is commonly found in cutlery and many household items, as well as in the marine and aerospace industries where exposure to water or high humidity is a concern. Blends of other elements—including nitrogen, aluminium, sulphur, nickel, copper, selenium, niobium, and molybdenum—can provide tailored requirements for various industries, and enhanced corrosion resistance can be provided through, for instance, higher levels of nickel or molybdenum. Steel use is widespread and is found in plate, sheet, tube, bar, and wire forms, and in castings and forgings.

White-hot steel pours like water from a 38-ton electric furnace at the Allegheny Ludlum Steel Corporation in Brackenridge, Pennsylvania, US.

The relative impact of oxidation on non-stainless-steel (left) and stainless-steel nuts, showing the level of rust and corrosion on the former.

Stainless steel

Stainless steel followed the discovery of chromium in the late 18th century, when it was first demonstrated at the French Academy. During the following decade into the early 19th century, its properties were tested and its resistance to acid made it an obvious choice for tableware. It was the Chrome Steel Works of Brooklyn that first employed stainless steel in the construction of bridges, and from this research came a variety of different stainless steels which readily found application in weapons. The different types of stainless steel are given a three-digit prefix number, and from these internationally recognized standards users select an appropriate elemental composition.

Aluminum

Aluminum is a light metal with a relatively low density and a high strength-to-weight ratio. It has good corrosion-resistant qualities, too. This makes it an ideal choice for a very wide range of manufactured products, such as containers, sports equipment, and parts for cars and vehicles in the transportation industry. It can come in a wide range of cast and wrought shapes, the latter involving it being mechanically worked into pre-selected forms. For most applications, there are two types:

1. Aluminum-silicon alloys, which are most applicable for casting.
2. Aluminum-copper or aluminum-magnesium alloys, which are best for mechanical working.

Magnesium

Magnesium has found wide application in the automotive and aircraft manufacturing industries, and its blending with aluminum in an alloy is probably the most common. It was used for German aircraft in World War I largely because of its strength and light weight, and later the Germans gave it the name *Elektron* for an alloy.

However, it has unpleasant properties in that it burns with an intensity and high temperature that can destroy entire aircraft. For instance, it was adopted as the crankcase for the engines used in the Boeing B-29 heavy bomber of World War II. Engine fires could ignite the crankcase, which would burn with a temperature in excess of 5,630°F (3,100°C). There were also instances where fires started in the engine itself ignited the magnesium and burned through the aircraft's entire wing like a blowtorch.

Copper alloy

The ductile nature of copper alloy makes it a favorite for reforming and for its resistance to fracture, while also being an excellent conductor of both heat and electricity. Because of these qualities, it is frequently used for electrical and electronic work, as well as plumbing. Not so long ago, it was used in telephone wires, and some buildings suffering from exceptionally poor broadband have discovered to their cost that the incoming line is copper! Copper has an added advantage in that if alloyed with zinc, it creates brass, which has a much greater strength than copper alone. It can be alloyed with tin, too, to make bronze, which in the modern world is the material of choice for engine bearings.

Superalloys

There is also a class of ***superalloys***, defined as alloyed materials capable of operating at a high fraction of their melt point without deforming, retaining high strength, good surface stability, and resistance to corrosion or oxidation.

Nickel superalloys have good heat resistance and high-temperature mechanical properties, and these are most commonly found as the material of choice for turbine blades in jet engines, where they can operate efficiently at temperatures of more than 1,800°F (1,000°C) and operate under very high stress levels.

In addition, there are several other superalloys, each with their own chemical advantages and available for use in extreme machine environments.

Titanium

Of special application in the modern world, titanium is a natural element, discovered in Cornwall at the end of the 18th century. With low density, high strength, and a silver color, it is corrosion-resistant in water and in several chemicals. For the design engineer, titanium has the highest strength-to-density ratio of any metal alloy, and in a natural, unalloyed state it is as strong as some steels, while being less dense.

Titanium can be alloyed with a variety of metals, including iron, aluminum, vanadium, and molybdenum to provide a class of materials suitable for the aerospace industry in particular. But titanium has also found favor in spectacle frames, where its strength and stiffness are an advantage, and in prosthetics, implants, and instruments for fine surgical or dental procedures. With a relatively high melt point—in excess of 3,000°F (1,650°C)—it is useful as a refractory metal, a class of metals with very high heat and wear resistance. It is available for exotic applications due in part to its non-magnetic properties, its poor conductivity only partially offset by its loss of strength above 800°F (430°C). But there are dangers in using titanium: as a powder or in the form of shavings, it is a fire hazard that, when only moderately heated in air, can cause an explosion.

Hypersonic superalloys

An early application of superalloys was the choice of Inconel for the skin of the North American X-15, a rocket-powered aircraft flown during the 1960s on hypersonic and extreme high-altitude test flights to the edge of space.

With a structure manufactured primarily from Inconel-X, materials for the hypersonic X-15 research aircraft were required to withstand external temperatures of more than 1,300°F (700°C).

CERAMICS AND POLYMERS

Ceramic Any hard, brittle, heat- and corrosion-resistant material typically of metallic elements combined with oxygen or carbon. Most are crystalline and poor conductors of electricity.
Polymer Naturally occurring or synthetic compound consisting of large molecules made up of a series of linked compounds.

While metalworking has been the cornerstone of the Industrial Revolution—arguably for the last several thousand years—the use of ceramics has followed a close second. As we remarked earlier, electrons are not free to move through ceramic materials in the way they are for metals. Ceramics are therefore poor conductors of heat and electricity. While bonds are very strong and stiff, exhibiting very high melting points, the nature of their atomic composition means their atoms are unable to slide past each other, which renders them liable to fracture and fail in a sometimes highly catastrophic way. But to the engineer, ceramics are a way of producing objects from inorganic, non-metallic materials through

Most superalloys used in modern jet engines contain mostly nickel with large proportions of chromium and cobalt. These were developed in Britain during the 1940s by Wiggins Alloys for the Whittle jet engine in the UK's first jet aircraft. Wiggins was acquired in 2014 by American company Ameriflight, and the application of their superalloys burgeoned into a wide range of extreme environments, including pressurized water reactors for nuclear power stations, natural gas processing plants, and a range of high-performance cars for F1 and NASCAR races.

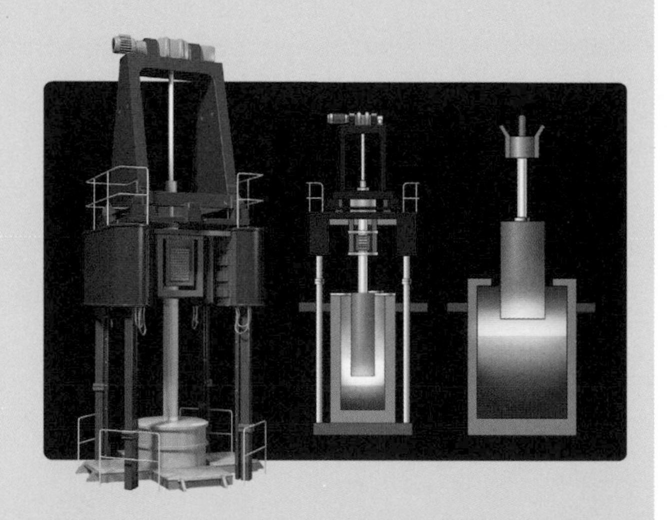

Refined steels and alloys are produced through a process of electro-flux remelting for particular use in the aerospace, nuclear, and military markets.

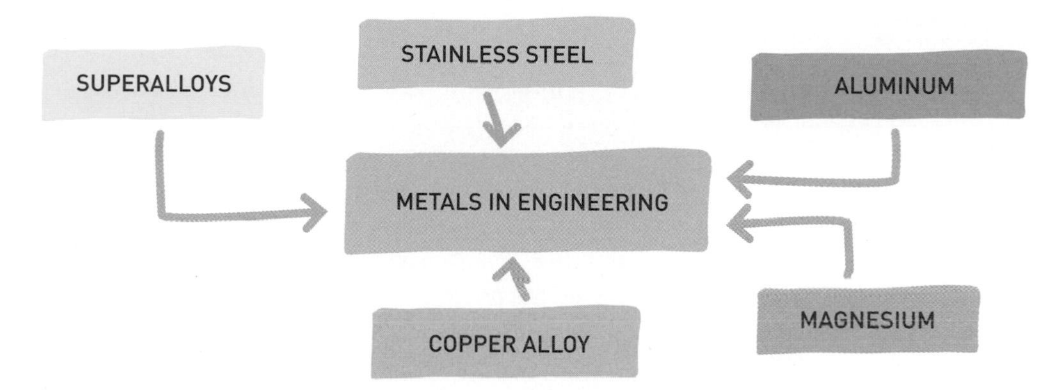

heat reactions from high-purity chemical solutions at lower temperatures. They provide the engineers with a solution to many tasks that cannot be carried out by metals or polymers, simply because they are heat-resistant.

For engineering purposes, ceramics are usually used in compression to avoid problems with tension and a threat to fracturing, but they have very specialized uses and limited applications. Nevertheless, there are still several categories where they are useful, one of which is ***alumina***, an aluminum oxide ceramic possessing hard yet brittle properties with poor electrical and thermal conductivity. This is good for ***thermal insulation*** as an ***oxidation barrier*** and it is useful in spark plugs for reciprocating engines, and has application with cutting tools and as an abrasive.

Silicon nitride ceramic has high thermal conductivity and toughness in comparison to other ceramic materials. It can be found in a range of cutting tools, in nozzles for grit blasters, in some turbochargers for motor vehicles, in selected-specification turbine blades, and it is also found in the shrouds for small jet engines.

Utilizing fibers of 5–10 microns, ***carbon fiber*** materials have a very high strength-to-weight ratio and a low ***coefficient of thermal expansion***. They are a natural favorite for reinforcing composite materials and are used in filters, electrodes, and antistatic devices and products. To produce them, carbon fibers are bonded in crystals with the majority aligned to the long axis of the fiber so that they are coincident with the crystal alignment. This produces the maximum high strength-to-volume ratio. They are more commonly found in combinations with other materials to form composites, and can be impregnated with resin to form a carbon-reinforced polymer with a high strength-to-weight ratio. When composited with graphite, they form reinforced ***carbon-carbon*** (RCC). By far the highest demand comes from the automotive and aerospace industries, but it is expensive and is applied sparingly and only where the high cost can be justified.

Synthesis of silicon nitride

Offering a highly specialized, albeit important, capability on the materials menu, silicon nitride is a compound of silica and nitrogen with a variety of thermal properties, and is relatively chemically inert. It is prepared by heating powdered silicon in a nitrogen environment to a temperature of 2,300–2,600°F (1,300–1,400°C) with weight increasing proportionate to the balance of silicon to nitrogen. Due to the potential dissociation to its two elements, this ceramic cannot be heated to more than 3,360°F (1,850°C)—well below its melting point. As a fascinating side note, only in the 1990s was it recognized in a natural state when silicon nitride was discovered in a meteorite where, as a mineral, it was named "nierite."

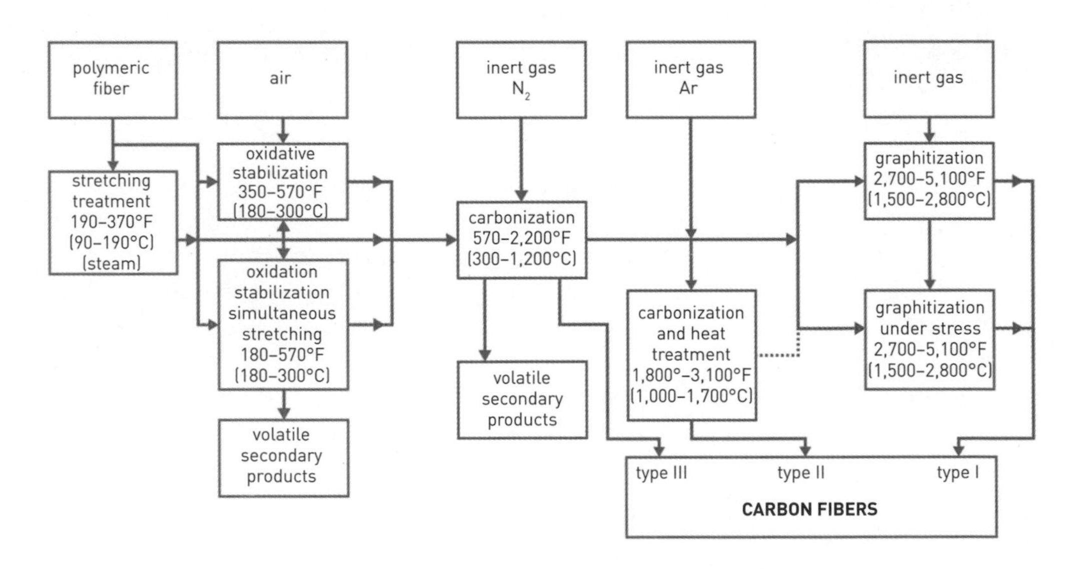

The preparation and processes in the production of carbon fiber from polymeric materials.

GLASS

Grouped with ceramics, ***glasses*** are a selection of amorphous materials that do not have a regular crystal structure and are usually produced when a viscous membrane cools rapidly.

Much as they are in a liquid, the atoms in glass are arranged in a pseudo-random fashion. Unlike metals, which can be formed into a glassy state, glasses usually comprise inorganic materials. It is because they have a non-crystalline structure that by their very nature light can be transmitted through them. But they are brittle materials because the bonding in glass is covalent, and the structure is vulnerable to fracture and stress. Glasses mainly consist of silicon dioxide, and specialist properties can be imported to improve durability, color and luster, and light absorption or transmission.

Glass has an advantage through its optical qualities both as a window and as a safety screen for sensitive instruments, but it can also be used as fibers for optical cables, insulators, and reinforcement for polymers.

COMPOSITES

Generally, ***composites*** consist of two or more constituents remaining separate within the completed structure.

Carbon fibers have high stiffness and strength-to-weight ratio but they can only be manufactured as fibers, and in that form they have limited value. However, as composites bonded together in a polymeric

Ceramics in space

One spectacular example of the use of ceramics in mechanical engineering was the thermal protection system for NASA's Space Shuttle, which, after leaving orbit, would fly to a landing on a conventional runway. To protect the Shuttle's orbiter from excessive heat—which built up on the underside as the orbiter used friction with the atmosphere to reduce its speed, where temperatures could increase to nearly 2,200°F (1,200°C)—it used ceramic tiles. Manufactured by the then Lockheed Missiles and Space Company and known as LI-900 (the type number originated from the bulk density of the material: 9 lb/ft^3 or 144.2 kg/m^3), the more than 20,000 tiles on the orbiter's underbody provided a heat-sink capacity and consisted of very pure quartz sand.

About 90 per cent of each tile was a void, hence the low density, and a black coating consisting of tetrasilicide and borosilicate glass was applied to the exterior surface area, apart from face bonding the tile to the aluminum skin of the orbiter. Individual tiles were shaped for a specific location on the orbiter, and on average they were 6 x 6 in (15 x 15 cm) in size. At a weight penalty, the LI-2200 tile was used on the relatively few areas where maximum emissivity was essential. These had a density of 22 lb/ft^3 or 352.4 kg/m^3.

Built around a Ford F-series chassis, the Plasan SandCat is a military vehicle protected by integrated composite armor.

matrix, it is possible to manufacture three-dimensional structures and in that application they have great value in mechanical engineering. In fact, they are enablers capable of supporting unusually strong structures for conditions and environments unaddressed by basic materials.

Yet none of this mixing of materials to enhance strength and performance is really new. In the most basic and crude analogy, it is akin in civil engineering to producing reinforced concrete structures with steel members to provide added strength.

Did you know?

Technically, ***wood*** is a composite material because the matrix contains fibers that show different properties and reactions in different directions. But generally, composites allow the properties to be "tuned" so as to meet a specific requirement.

Types of glass

- Variations include soda-lime glass, where sodium oxide and calcium oxide are added to silica to produce a glass with a low melting point that can be easily formed into windows, bottles, and many different types of light bulb.
- When boron oxide is added to make heat-resistant and low-expansion material, the resulting borosilicate glass produces a specific type of cookware, for instance Pyrex, which is also used for laboratory equipment.
- For its good fiber-forming qualities, E-glass is formed by adding aluminum oxide, calcium oxide, and magnesium oxide to the silica base. This is ideal for reinforcing glass-fiber-reinforced polymer composites, and is generally referred to as ***fiberglass***.

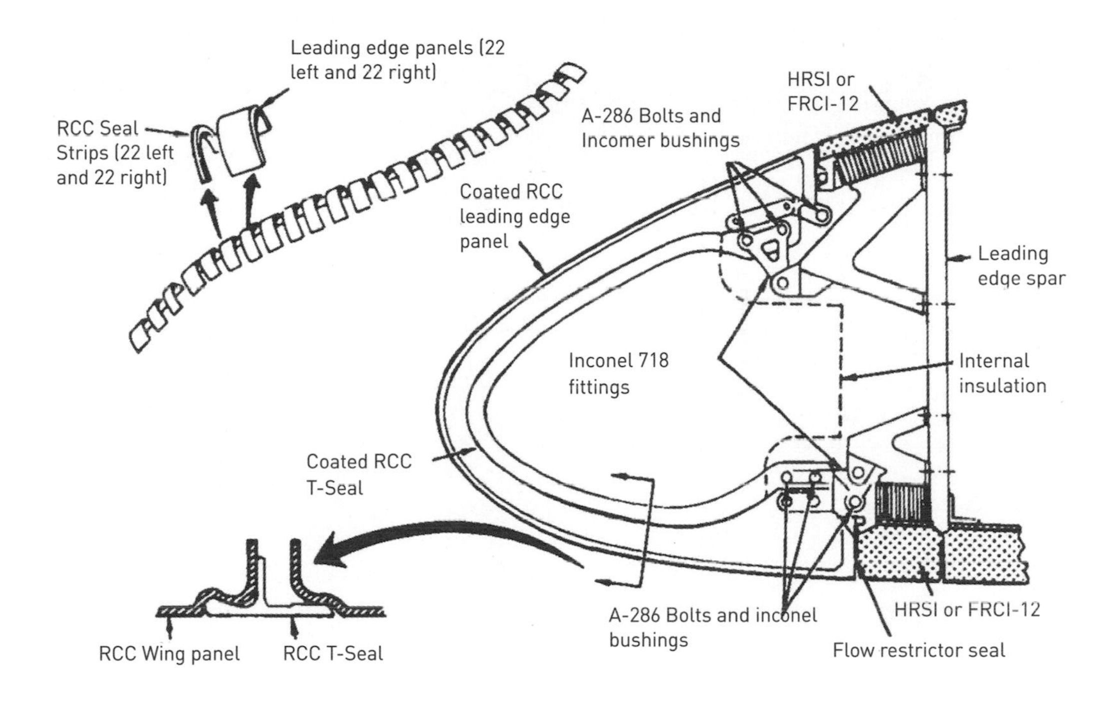

The leading edge of the Shuttle orbiter delta wing consisted of 22 reinforced carbon-carbon (RCC) composite materials to protect the structure from intense heat.

MATERIAL FAILINGS

Across the range of available materials, metals, ceramics, glasses, polymers, and composites have underpinned much of the technology and mechanical engineering in the modern world. But the interplay of different materials for specific purposes can sometimes be compromised by poor design and ineffective precautions. Chapter 8 examines risk and the responsibility of the design engineer in greater depth, but there is no doubt that materials and their integration into working machines is one of the biggest challenges in 21st-century engineering. In a chilling reminder of just how different the qualities of these materials are, and how easy it is to construct a sequence of failure with catastrophic results, one of two Shuttle disasters in its 135-mission history occurred on February 1, 2003 when the orbiter *Columbia* was returning from space and broke apart due to a failed material.

As we have seen, the Shuttle was able to perform its mission in part due to the use of composite materials as well as the ceramic tiles used to clad the exterior. The re-entry trajectory required the orbiter to experience temperatures in hot

The orbiter's RCC

The RCC on the orbiter consisted of a laminated composite made from carbon fibers impregnated with a phenolic resin, which, after being cured in a high-temperature autoclave, was pyrolized to convert the resin to pure carbon. A further process impregnated the material with a ***furfuryl alcohol*** in a vacuum chamber before it was again cured and pyrolized to convert the alcohol into carbon. The exterior face of the RCC was coated with a silicon carbide to prevent oxidation.

zones that exceeded the capability of the tiles. To protect especially hot areas—for example,trhe leading edge of the double-delta wing and the nose cone, which experienced temperatures up to 2,750°F (1,510°C)—a RCC material was applied; each wing had 22 RCC panels between ¼ and ½ in (6 mm–13 mm) thick.

A major downside of the RCC was that it could not take impact shock and would fracture if struck by a heavy object, or by a small object traveling at high velocity. This is exactly what happened to the orbiter. During ascent, shortly after lift-off, a small piece of insulation from the external tank, which provided cryogenic propellants to the three main rocket motors, came loose at an altitude of 12 miles (20 km) and a speed of Mach 2.46 and struck one of the 22 RCC panels on the leading edge of the port wing.

Manufactured from a ***polyisocyanurate*** (a thermoset plastic), the insulation was sprayed on after tank fabrication so as to maintain cryogenic temperatures for the propellants. Because the insulation material was ***anisotropic***—possessing different properties in different directions—and because of the velocity at which the insulation struck the panel, it shattered it. On return to Earth, heat entered the void on the port wing and began to melt the unprotected aluminum structure, causing the orbiter to break up across the southern states of the US and killing all seven crew members on board.

An impact test on an RCC panel showing the vulnerability to even small, low-mass materials striking the surface at high velocity—the cause of the total loss of the Shuttle orbiter Columbia *in 2003.*

Chapter Six

KEEPING THE MACHINES MOVING

Heat Engines—Petrol Engines—Diesel Engines—Two-stroke Engines—Aircraft Engines—Jet Engines—Engines Evolve

HEAT ENGINES

So far, we have seen how the use of forces and principles of chemistry and thermodynamics have shaped the modern world of mechanical engineering. We have seen the broad array of materials, many of them natural elements, from which the engineer can choose those with specific characteristics or capabilities for particular functions. We have also seen how materials not found in nature have been produced for specific applications. It is now time to see how these operating principles and materials when brought together can make the world of machines move in a safe, productive, and interactive way, in particular the internal combustion engine and reaction engines.

Mechanical engineering is an assembly of separate specialist categories. In addition to those we have dealt with so far, other important categories include mathematics, fuels (and associated applications in combustion processes), control theory, hydraulics, pneumatics, and CAD/CAM. A working example of how those other disciplines come into play is the ***heat engine***: a system that converts heat or thermal energy into mechanical energy, which is then used to perform work. It achieves that by bringing a working substance from a higher-state temperature to a lower-state temperature. Specifically, a heat engine is distinguished by its efficiency being limited by the ***maximum thermal efficiency*** it can obtain.

The Carnot theorem

The limiting efficiency for heat engines depends upon the temperatures of the hot and the cold reservoirs in the system, and this is known as the Carnot theorem after Nicolas Léonard Sadi Carnot (1796–1832), who developed it in 1824. A defining factor is the gap between the hot source and the cold sink; the greater that is, the better the thermal efficiency of the cycle.

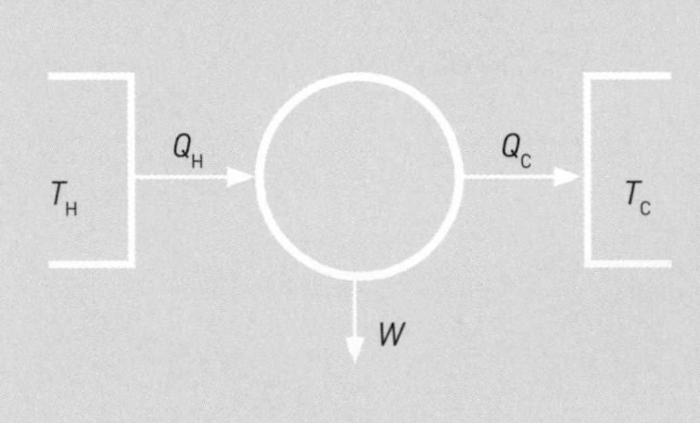

A diagram of the operating principle of the Carnot engine—the internal combustion engine—where heat Q_H flows from a high temperature T_H furnace through the fluid of the working substance and the remaining heat Q_C flows into the cold sink T_C, thus forcing the working substance to perform mechanical work W on the surroundings, via cycles of contractions and expansions.

PETROL ENGINES

Today, the most frequently encountered form of heat engine, and the optimal one for describing all the interconnected categories identified above, is the ***internal combustion engine*** as applied to ***petrol-fueled motor vehicles***. These are ***phase-change engines*** because the working fluids are gases and liquids; the engine converts the working fluids between phases (liquid to gas or gas to liquid) into effectively useful work from ***fluid expansion*** or ***compression***.

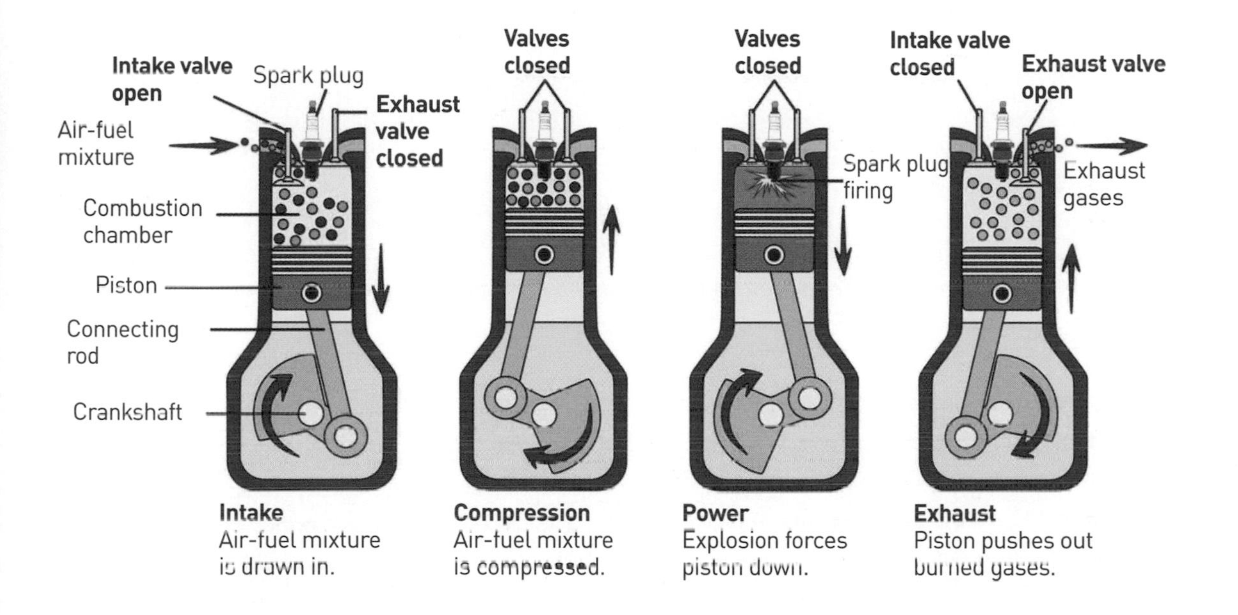

The operating principle of the international combustion engine through intake, compression, power (combustion), and exhaust strokes—the four-stroke engine.

The Otto cycle

The conventional four-stroke engine was designed during the mid-19th century and patented by Alphonse Beau de Rochas (1815–93) in 1861. However, the first working engine of this type was designed and built in 1876 by Nicolaus Otto (1832–91), hence it is his name that has been given to the standard, four-stroke combustion engine employing spark plugs for ignition—also known as the Otto cycle.

The four-stroke engine

The very name of the system implies four separate "strokes" of a piston moving from the bottom to the top of a cylinder: the ***intake stroke***, ***compression stroke***, ***power stroke,*** and ***exhaust stroke***. The intake stroke draws in air and fuel for detonation under compression in the combustion chamber, followed by the opening of the exhaust valve for evacuating the exhausted gases through the exhaust manifold, while the piston delivers force to the ***crankshaft***, which conducts a rotary motion, and from there to the ***driveshaft***. But that is a simplified overview. The piston is attached to the crankshaft by a connecting rod, with the "little end" inside the piston and the "big end" connecting the rod to the crankshaft via an offset from its center line. As the piston moves up and down, it rotates the crankshaft. It is the regular firing of the cylinders that creates the power to keep the crankshaft rotating with a continuous torque in a steady fashion.

In the operating system, an overhead camshaft is required to operate the inlet and exhaust valves, or there may be two camshafts, one each for the inlet and exhaust valves respectively. These are referred to as ***single*** or ***double overhead camshafts*** (SOHC or DOHC). The camshaft rotates and provides a ***cam train*** aligned with the inlet and exhaust valves, controlling them for the correct sequence for stable combustion. The rotation of the camshaft above the cylinder is directly controlled by the crankshaft below it, via belts and pulleys.

While early four-stroke automobile engines used a carburettor to mix fuel and air, Mercedes-Benz built the OM 352, a direct-injection engine that electronically controls the injection of a mixture into the combustion chamber.

Multi-cylinder engines

On ***multi-cylinder engines***, the firing sequence is controlled by the angular alignment of the cams as they determine the order in which combustion takes place. The precise timing for spark plug ignition is usually adjustable, a facility that is now electronic rather than mechanical. In some high-performance engines, each cylinder has two inlet and two exhaust valves, and this is now increasingly common in road cars, too.

Cylinder arrangements

The arrangement of the cylinders in a typical multi-cylinder engine is commonly in-line or in a V configuration. For instance, a four-cylinder engine would have the cylinders aligned with the long axis of the vehicle, whereas a transverse engine would have them at 90 degrees to the axis of travel. A V engine of perhaps six, eight, or 12 cylinders would have them divided into equal numbers set at a V of varying angular displacement. In the case of a V engine, there would be one or two camshafts per cylinder bank, the entire power system effectively operating as two separate banks of cylinders united by a common crankshaft.

It is the firing sequence in a multi-cylinder engine that ensures that each cylinder fires in a specific sequence in relation to its adjacent cylinder, thus maintaining a continuous torque in a steady and even fashion for minimal vibration. It is important that the firing of all cylinders does not occur simultaneously because if all the cylinders fired at the same time, it would set up a ***resonant oscillation***, adding stress through the system.

Diesel vs. petrol

In comparing the two types of engine, the Otto cycle is the more efficient but overall the diesel wins out because it is able to operate at higher compression ratios, and self-ignition is, theoretically, more efficient. But the advantages and disadvantages of each are finely tuned and highly selective so that there is no clear-cut overall advantage in either, selection being according to the type of application to which the engine is assigned.

DIESEL ENGINES

The ***diesel*** or ***compression-ignition engine*** is a ***reciprocating internal combustion engine*** whereby fuel is ignited by heat generated during compression of air to which an elevated-temperature mix of additional fuel is added, obviating the

The origins of the two-stroke engine

Developed as a concept in 1881 by a Scottish engineer, Dugald Clerk (1854–1932), the two-stroke design was patented in 1881. The early version had a separate charging cylinder, an element that had been introduced by Clerk based on an idea by Joseph Day (1855–1946). However, it is to Englishman Alfred Angas Scott (1875–1923) that credit goes for the first practical two-stroke engine, which he installed in a two-cylinder water-cooled motorcycle in 1908.

need for a spark plug. It works by increasing the inside air temperature to a level at which atomized fuel spontaneously ignites. Because there is no physical device to ignite the fuel, the injector sits in the position otherwise occupied by the spark plug in a four-stroke engine.

The most common form of fuel for diesel engines is a fractional distillate of petrol, but it can be biomass, gas, or biodiesel types. Synthetic substitutes have been around since the late 19th century and can be used equally in either a diesel or a petrol engine.

TWO-STROKE ENGINES

As an alternative to the four-stroke Otto design, the two-stroke engine completes a single power cycle with one up and one down movement of the piston for a single revolution of the crankshaft, rather than four piston strokes for two crankshaft revolutions. It achieves this by combining the end

A two-stroke radial engine built by Nordberg, applied to flood control at Lake Okeechobee, USA.

Attributed to Rudolf Diesel (1858–1913), the origin of the diesel engine dates to the late 19th century, as represented here by a Langen & Wolff motor of 1898.

of the combustion stroke and the start of the compression stroke simultaneously with the intake and exhaust functions.

These engines have a higher power-to-weight ratio, with the power available in a narrow band of rotational speeds. Two-strokes also have a greatly reduced number of moving parts. Two-stroke engines are lubricated by a mixture of petrol and oil in a ratio of 1:50 so that the oil will form emissions that result in more toxic substances than a four-stroke engine, the operational cycle capable of releasing unburned fuel vapor while the high temperatures can release nitrous oxides.

AIRCRAFT ENGINES

The four-stroke engine was readily adopted for the motor car and associated vehicles, remaining as the mainstay for road transportation throughout the 20th century and beyond. There are alternatives, of course, and we will look at electric power systems in Chapter 12. In addition, the reciprocating engine was used to propel heavier-than-air flying machines from the beginning of the 20th century.

In the case of its application to flying machines, the crankshaft of the reciprocating engine drove a driveshaft attached to a propeller rather than to a differential and wheels on a rod vehicle. As with so many technologies, war brought the biggest advances, and during World War I reciprocating engines for aircraft evolved in terms of both engineering design and power.

Types of aircraft engines

The conventional in-line engine was initially popular, rapidly joined by ***radial*** and ***rotary aero-engines***. Usually supporting an odd number of fixed cylinders spaced at equal intervals around a central crankshaft, the radial engine was air-cooled rather than water-cooled, and became a firm favorite with pre-war aircraft designers. The radial engine evolved in the late 1810s into the rotary engine, in which the cylinders rotated round the crankshaft, and this type rapidly found favor during World War II. The difference between the two was that for the rotary engine the propeller was bolted directly to the engine itself, and the crankshaft to the airframe.

A Wright R-2600 radial engine (its number derived from the output in imperial horsepower in a system originating with James Watt) on a World War II B-25 Mitchell bomber.

The origins of the jet engine
Of course, it could be said that the first "jet" engine was Hero's aeolipile of nearly 2,000 years ago, but that would be inaccurate in that it was not capable of linear motion under reaction from a directed, combusted fuel. Throughout the late 19th and early 20th centuries, several pseudo-jet designs and configurations were theorized, designed, and tested, but not until the 1930s did a working jet engine fire up and produce a design capable of propelling an aircraft in sustained flight.

Over the next several decades, radial engines became popular while rotary engines fell out of favor. That said, as variations of a basic principle, in-line, radial, and rotary engines all played their part in the development of powered flight. For instance, the in-line engine was put to use replacing steam engines in ships of all kinds during the 20th century.

JET ENGINES

Another application of the heat engine, the ***air-breathing jet engine*** is a comparatively recent invention and uses reactive forces defined by Newton's third law (see page 35). It operates by generating thrust derived from combustion of a fuel-air mixture—a process that focuses an exhaust outlet in one direction and by that action incurs a reaction that drives the engine, and the aircraft to which it is attached, in the diametrically opposite direction. In this category can be included the rocket motor, which paradoxically predates the jet engine, which conducts combustion of a fuel and an oxidizer, both of which are integral to the rocket to which the motor is attached.

There are many different types of jet engine, but they all operate on the same basic principle. The most important is the ***convergent-divergent nozzle***, which turns internal pressure into high-velocity kinetic energy, a process whereby pressure and temperature remain the same through the nozzle but the static value drops as the gas is accelerated. After combustion, the velocity of the gas stream entering the nozzle is less than half the speed of sound (Mach 0.4), which is essential for minimizing the loss of pressure in the duct from the combustion chamber to the nozzle itself. The standard convergent nozzle is capable of accelerating the gas to a localized supersonic speed, but the convergent-divergent profile of the nozzle is what produces much greater exhaust velocities.

Centrifugal-flow engines

Several countries were involved in the development of jet propulsion, the most notable being Germany and Britain, where RAF pilot Frank Whittle (1907–96)

A de Havilland Goblin II using the centrifugal principle for jet propulsion originally developed by Sir Frank Whittle.

pursued a personal campaign to get funds for development of his ***centrifugal-flow design***. As a cadet at RAF Cranwell in 1928, Whittle had several ideas about reaction propulsion and conceived the marriage of gas turbines and jet propulsion, registering a patent for a turbojet on January 16, 1930.

In the gas turbine, air is moved through a compressor to achieve higher pressure with added energy provided through fuel sprayed into the compressed air, igniting it for a high-temperature flow to drive a turbine, where it is used for work, exhausted gases being redundant. Whittle's idea was to take the redundant gas and repurpose it for producing direct thrust through a convergent-divergent nozzle, turning the gas turbine into a ***reaction engine*** and using it to propel a flying machine.

Supercharger vs turbocharger
As a side note, a ***supercharger*** should not be confused with a ***turbocharger***. The latter is not mechanically driven by the engine like the supercharger, but is instead provided with a small turbine to deliver highly compressed air to the combustion process.

Whittle's jet engine used the centrifugal-flow concept, whereby air was sucked in from the intake by an ***impeller***, the compressed air then being thrown out to a ***toroidal ring*** of 16 combustion chambers where fuel was injected. The combusted gas products were passed through a turbine that drove the impeller before passing through the convergent nozzle, which accelerated the flow. This was a logical development of the centrifugal supercharger used on motor vehicles, but it had the disadvantage that the impeller (or compressor) necessarily had a very large frontal area to achieve the amount of air required for combustion. The ***supercharger*** was simple in principle, and increased the pressure or density of air delivered for combustion in a conventional reciprocating engine by providing more oxygen for each intake cycle, thus allowing it to burn more fuel, produce more work, and deliver more power.

Axial-flow engines

Centrifugal-flow was only one of two possible concepts for a jet engine, the other being an axial-flow engine, which in simple terms is a turbine in reverse, a concept favored by the Germans. Here, air entering at the front is moved to the rear of the engine by a fan stage, where it is presented to a series of static ducts known as stators. Because this configuration is weaker than the centrifugal-flow concept, the number of fans and stators is necessarily larger. These are placed in series so as to maintain the minimum

Jet engines in action

During the late 1930s and early 1940s, German engine designers opted for the axial-flow concept and applied it to achieve the world's first official flight of a jet aircraft, the Heinkel He 178, which took place on August 27, 1939 (although a short hop had been performed a few days previously). Over the next five years, a small number of jet prototypes flew, until the world's first operational jet fighter, the Messerschmitt Me 262, became operational towards the end of 1944, quickly followed by the first jet bomber, the Arado Ar 234.

Elsewhere, chronologically, the Italians were the second nationality to fly a jet aircraft. The Caproni Campini CC.2 was powered by an Isotta Fraschini radial engine driving a ducted fan compressor where air passed through a variable nozzle exit pipe into which fuel could be injected for combustion. It took to the air exactly one year to the day after the He 178. The world's third aircraft to fly with a jet engine, the British Gloster E.28/39, got airborne from RAF Hucclecote in England on April 8, 1941, powered by a Whittle-designed engine, the River W.2B.

frontal area of the engine—a major advantage with the axial-flow design. The design is clever. The energy of the fluid (air) increases as it flows through the various compressor fans because of the action of the rotor blades, which exert a torque on it, while the stators slow the air and convert the circumferential flow into pressure. Because of this, axial-flow engines produce a continuous flow of compressed gas with high efficiency and a large mass flow rate.

Jet engines diversify

The early engines were turbojets, accelerating air to supersonic speeds. Over time, the development of the jet engine has progressed beyond the expectations of many engineers involved in its inception, and has spawned several applications. These include the ***turboprop engine***, where the exhausted gases are used to drive a propeller shaft, diverting only some of the energy into a conventional propeller. The work of the propeller is supplemented by a fraction of the expended fuel-air mixture in an exhaust flow, adding thrust to the work conducted by the propeller itself and increasing the overall efficiency of the system. Turboprop engines are ideal for subsonic aircraft, where fuel efficiency is an advantage.

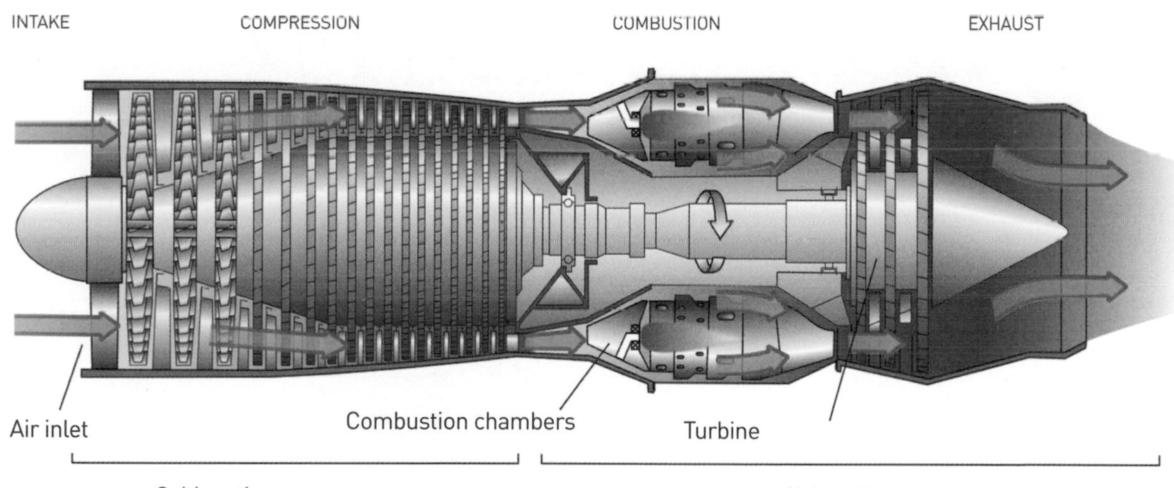

The principle of the modern turbojet engine. Just as with a four-stroke reciprocating engine, the turbojet—also a heat engine—has an intake, compression, combustion, and exhaust cycle, in this case the exhaust producing the work to induce a reaction. Hence, like a rocket motor, this is a reaction engine.

Specific fuel consumption

At this point, we can introduce the very important parameter of ***specific fuel consumption***, or ***sfc***, which is a direct measure of the efficiency of the fuel with respect to thrust produced.

The sfc will vary according to the condition of the atmosphere, the altitude, and the throttle setting used at any point in the calculation. It can apply to rocket motors, too, but in the specific instance of a jet engine, altitude is very important because it is approximately proportional to air speed and the ground speed is proportional to air speed. Because work done is force times distance, mechanical power is force times speed.

While sfc is used as a measure of fuel efficiency, it can also be used to compare different engines flying at different speeds, by dividing it by the speed. One example of this was the Concorde supersonic airliner, which cruised at a nominal speed of 1,354 miles/hr (2,179 km/hr) and had a nominal sfc of 0.452 kg/(kgf•h), which shows that the four Olympus engines transferred 17.9 MJ/kg, equivalent to an sfc of 0.2268 kg/(kgf•h) for a subsonic aircraft flying at 570 miles/hr (917 km/hr), which is about 42 per cent of the speed of Concorde. This demonstrates that the Olympus 593 engine was the most efficient turbojet engine ever placed in production—and in fact was a derivative of the engine that powered the Avro Vulcan V-bomber.

Calculating sfc

Sfc can be calculated as fuel in grams/second per unit of thrust (kilonewtons) and is derived by the consumption of fuel divided by the thrust output. Because the power is thrust times speed, $\eta = V / (SFC \times h)$, where V is speed and h is energy content per unit mass of fuel.

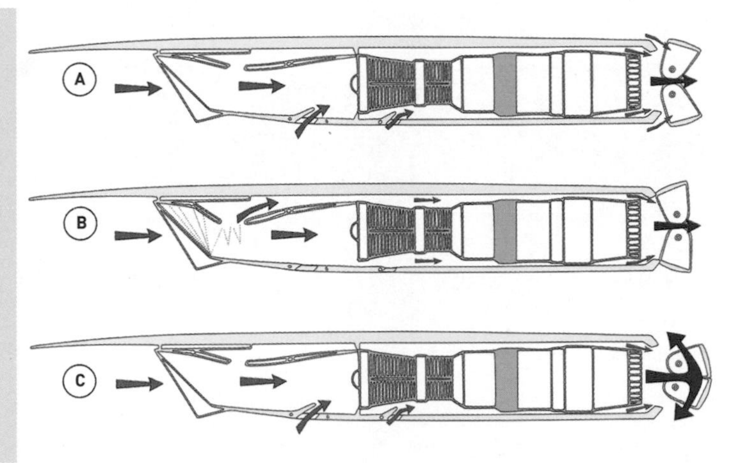

Integrated engineering of jet engines requires several mathematical integrations involving airflow, gas dynamics, and propulsion, as shown here for an Olympus 593 on Concorde. Respective intake ramp positions are shown for take-off at subsonic speed (A), supersonic cruise around Mach 2 (B), and reverse thrust to help decelerate the aircraft after touchdown. This adjusts the velocity and volume of air entering the engine at various stages of flight.

Seen here at the Aerospace Bristol Museum, England, an Olympus 593 turbojet with reheat capability of the engine family that powered the subsonic Avro Vulcan V-bomber and the supersonic Concorde commercial transport.

Reheat or afterburner

The Concorde engine also had ***reheat***—or ***afterburner*** as it is known in popular parlance—a means of increasing thrust by increasing the velocity of the exhaust while adding a small amount to the expelled mass, both of which significantly increase the power of the engine.

Because the temperature of the combustor is limited to about 3,700°F (2,040°C) by the materials available for a balance between heat and the durability of the structure, fuel is burned at about 3,850 kg/h using a relatively small amount of the air entering the engine. To bring the temperature down to a turbine entry level of about 1,560°F (850°C), the combusted products have to be diluted with air from the compressor. This provides the turbine with a safe level for long life, but puts a limit on the thrust level generated. If all the oxygen available were burned in the engine, temperatures would rise to the same level as those in the combustor and completely destroy the engine. However, the excess air remaining at a fuel/air ratio of 0.014 is sufficiently large to reignite it after it has passed through the turbine and into the exhaust tube, raising the temperature in free air to 750°F (1,390°C), which accelerates the gas on exit and slightly raises the mass flow.

Applications of afterburner
Engines with reheat are described as "wet" when in afterburner or dry when not, and the relevant thrust levels are always quoted for respective operating modes. Since afterburner requires prodigious amounts of fuel and cannot be used for extensive periods without compromising the range of the aircraft, it is an operational choice and is only adopted as a design requirement when a significant burst of speed is required in the aircraft's overall performance, usually for military aircraft.

Turbofan

For less demanding performance requirements, and where sfc and overall fuel efficiency is a driving criterion for a commercial success, the engineer has the option of a ***turbofan engine***. But to remind ourselves, thrust depends upon the velocity of the exhaust gas and the mass of the gas ejected. The pure jet or turbojet design favors the velocity of the exhausted gas in precedence over mass, whereas a turbofan engine prioritizes mass over velocity. The turbofan differs only in that it uses a standard turbojet to drive a ducted fan to move more air across the operating cycle of the engine and thereby contribute to thrust. It is the concept of choice for most aircraft today and is the power plant of necessity for the modern airliner, as well as for small and medium-size engines. It is in this sector of the aircraft engine market that engineers are pressed to make the greatest improvements in sfc and lower operating costs.

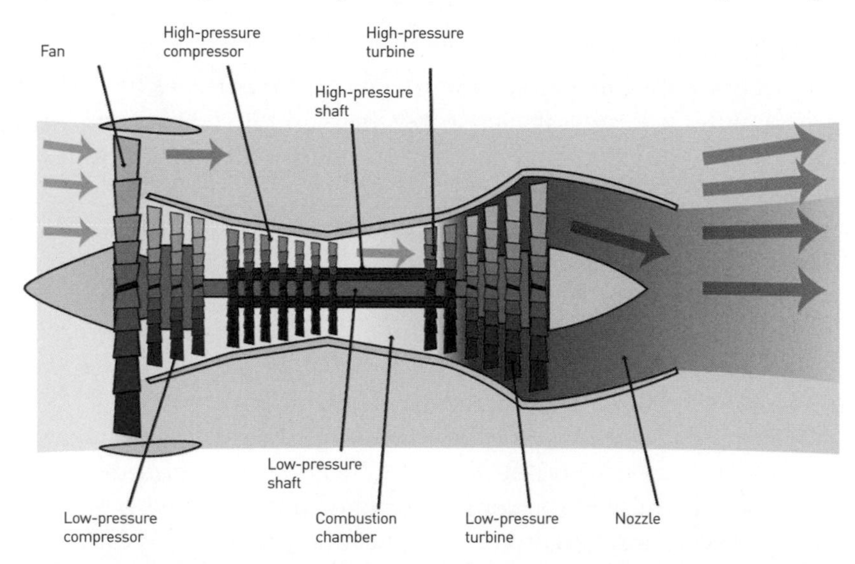

The operating principle of the turbofan engine, where air is drawn into low- and high-pressure compressors to the combustion stage, driving a low-pressure turbine before exhausting combusted products.

Quantifying and balancing performance

The performance advantage of an engine for a modern airliner can be gauged by the thrust transmitted to the airframe given by the mass flow of air passing through the engine times the increase in the speed of that air. Given that air entering the inlet at flight speed is V_{flight} and the air ejected at the rear is a nozzle speed of V_{jet}, if the mass flow is W, the thrust F is delivered through the equation producing the momentum thrust: $F = W(V_{jet} - V_{flight})$. To increase the mass flow W, the frontal area of the engine must be increased, which will also result in it being heavier and increase drag. But raising the V_{jet} increases fuel consumption for a given thrust.

The balance between the two is the function of the design engineer given the performance required. However, this is only valid for an engine that is not choked and where V_{jet} is less than Mach 1. When V_{jet} is fixed at Mach 1, pressure thrust is added to calculate F, as in the equation: $F = W(V_{jet} - V_{flight}) + A(P_{exit} - P_{inlet})$, where A is the jet exit area at the exhaust nozzle, P_{exit} is the static pressure at the nozzle inlet and P_{inlet} is the static pressure at the engine inlet. With V_{jet} at Mach 1, the new term for pressure thrust allows thrust to be increased by raising P_{exit}, which is achieved by raising the total pressure in the jet exhaust pipe. With V_{jet} fixed at the speed of sound, and the engine running hotter, the speed of sound can be increased so that V_{jet} is raised and momentum thrust increased proportionately.

The pros and cons of the gas turbine

Overall, the gas turbine is an ideal demonstration of the basic laws of thermodynamics. It shows that the power required for a given pressure ratio is proportional to the entry temperature. The turbine entry temperature can be five times that of the compressor entry temperature. Because of this, the turbine requires a much lower expansion ratio to drive the compressor than the compressor requires to do its work. The difference between the two is therefore available to produce thrust. Compared to the reciprocating engine, the core of the gas turbine can be 20 times as powerful because the continuous cycle and the large flow path of the turbine ingests about 70 times the volume of air over the same period. It should be noted, however, that although it could be inferred that this would allow 70 times the quantity of fuel to be burned, only about one-third of the air is used for combustion and the energy thus released is only about 23 times that of a piston engine—

which utilizes almost all the air for propulsion. Nevertheless, it's a significant increase!

On the downside, the clear performance advantages of the jet engine come at a price—that of complexity and cost, in materials and in fatigue wear (about which more later). Most costly is the inventory of materials required for a turbojet engine, notably the combustor and turbine materials, and while temperature and pressure in a reciprocating engine can be as high as those in a turbojet, these last only for very brief periods of each cycle and the materials needed for the former are not as expensive.

ENGINES EVOLVE

Whether for motive power or to drive static machines, mechanical heat engines operating with hydrocarbon fuels have transformed the world of engineering, and none have been more dramatic than the jet engine and the rocket motor. The enormous progress made in engine design for the modern airliner have significantly reduced the weight of the engine per unit power output, made major reductions in sfc as related earlier, and created an air transport industry that has revolutionized the mobility of goods and the movement of people around the world, and stimulated an industry that employs more people than any other in the world today.

Yet for all the advantages presented by modern air transport, this method of transport is doubtless a major consumer of hydrocarbon fuels. Fortunately, mechanical engineering is nothing if not used to rising to challenges, and as we will see in Chapter 13, unsustainable dependency on the airline industry is being addressed by the engineer in ways that could bring the next revolution in transport—the carbon-neutral airplane.

Chapter Seven

ENGINEERING THE SYSTEMS

Systems Management and Engineering—Systemic Failure—Keeping Control—Systems Engineering—Programme Evaluation and Review Technique

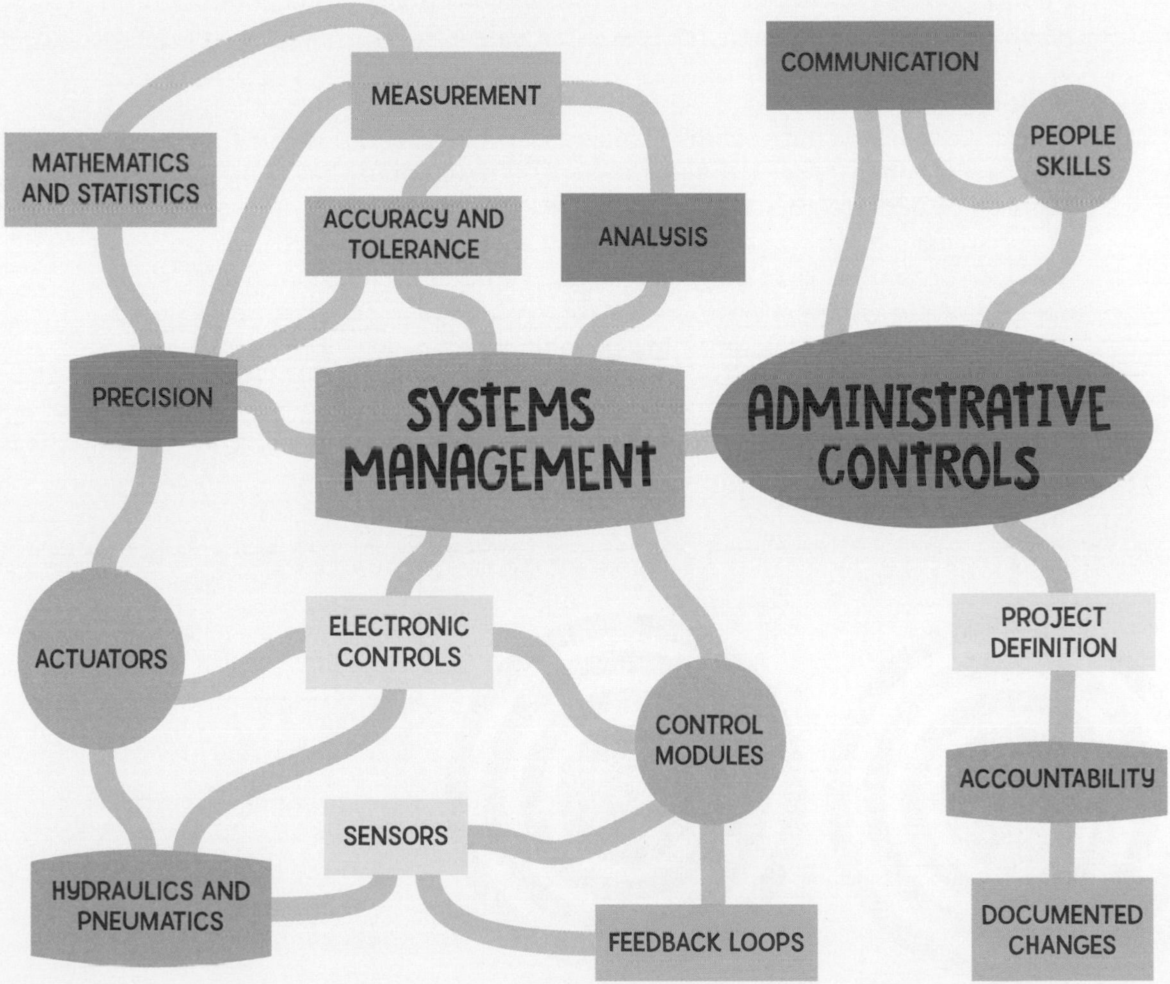

MEASUREMENT
COMMUNICATION
PEOPLE SKILLS
MATHEMATICS AND STATISTICS
ACCURACY AND TOLERANCE
ANALYSIS
PRECISION
SYSTEMS MANAGEMENT
ADMINISTRATIVE CONTROLS
ACTUATORS
ELECTRONIC CONTROLS
PROJECT DEFINITION
CONTROL MODULES
ACCOUNTABILITY
SENSORS
HYDRAULICS AND PNEUMATICS
FEEDBACK LOOPS
DOCUMENTED CHANGES

SYSTEMS MANAGEMENT AND ENGINEERING

So far, we have explored the ways in which forces and basic laws of physics and chemistry have underpinned mechanical engineering in the broadest sense—and we have looked at a few examples in moving machines on land and in the air to see how they are applied and manipulated. But there is more to these machines than their static or moving operation; there are ancillary services that each requires to maintain stresses, forces, temperatures, and pressures within acceptable limits, and there is also a need to monitor the condition of these engines to ensure that safety features can be activated to prevent catastrophic damage. All this is about managing the systems. We will also look at systems engineering as a science in itself—a means of controlling design, development, production, and operation.

Defining parameters

In order to understand systems management, we need to look at how a wide range of parameters are defined and kept in check, and that comes down to just two things: accuracy and precision.

In engineering terms, accuracy defines in numbers how close the measurement of a tolerance or a set of parameters comes to a particular value. Precision is about how close those measurements are to each other. Yet accuracy is not just one objective (albeit sometimes an essential requirement)—it is a description of systemic errors and a measure of statistical bias. And that is a product of the difference between a result and a true value, which the International Standards Organization (ISO) calls a "trueness" factor.

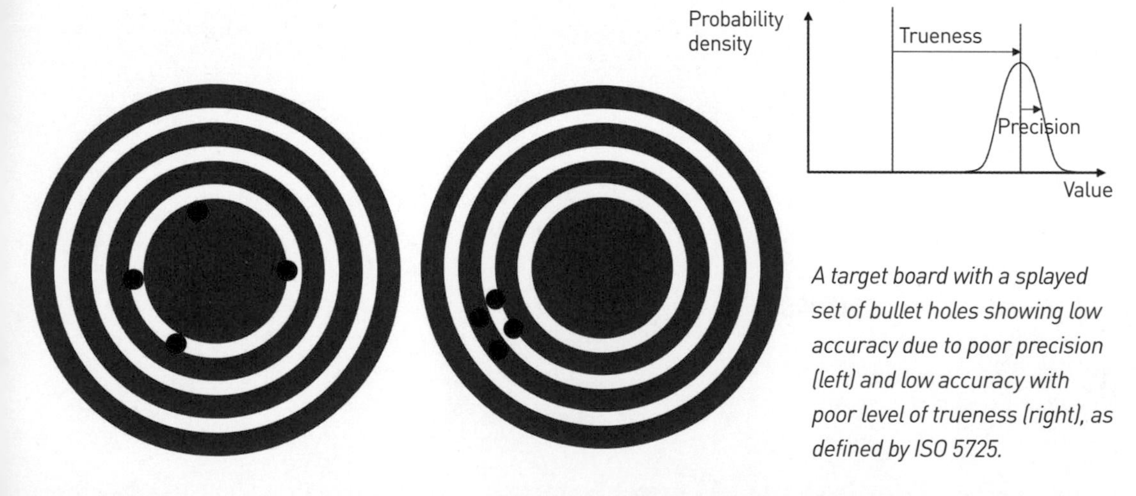

A target board with a splayed set of bullet holes showing low accuracy due to poor precision (left) and low accuracy with poor level of trueness (right), as defined by ISO 5725.

When talking about accuracy, in fine-scale definition, the trueness is the absolute to be sought—a number on a graph, an exact measurement to an infinite value—while the result is the product of a measurement, or an observation, for how the actual measurement matches the trueness factor. We can get side-tracked into statistics here, and we will deal with that field as it relates to mechanical engineering later, but in engineering as pertaining to measurement, a ***value can be both accurate and imprecise and imprecise but not accurate***. It can be neither of these or both. This can best be understood by an example. If on a number of occasions a measurement is taken and at each one there is a drift in the result, by expanding the size of the sample run, the precision will increase as a percentage error in the total spread; the sampling scale changes but the measurements have a much less evident drift from the trueness. It is this misunderstanding on the part of a shockingly large number of engineers that results in catastrophic disasters.

Types of error

Precision is the sum total of random errors made in the measurements, but a definition of the measurement itself is vital for qualifying the value of the measurement. Random errors are a measure of the difference between the value of a quantity and its true value. In engineering, an error is a mistake, but in statistical analysis—of failure probability perhaps—an error is not classed as a mistake. Errors in measurements can be divided into random errors and systemic errors. As the definition states, random errors are where there are inconsistencies in a sequence of measurements repeated under the same conditions and the same quantity or actuarial spread. Systemic errors are made when inaccuracies are observed, not by chance but by the introduction of an unexpected intrusion by either an inherent failure within the measured system or by the introduction of an unexpected parameter that destabilizes the device being measured.

We will carry these requirements further when we look at the overall concept of reliability and fault analysis, but a couple of examples illustrate the way poor performance of a total system can be introduced by the deliberate intervention of an external design requirement. The most obvious example is that inherent within the motor industry, where commercial success depends upon sales—to either wholesale or retail customers—and where engineers are required to design commercial products into systems that reduce the accuracy of a completely integrated system and the precision designed in for its function.

The control room for a Distributed Control System (DCS) where plant information and controls are displayed on computer graphics screens. The operators are seated and view and control any part of the process from their screens, while retaining a plant overview. Previously, such rooms were centralized with no integration between systems.

SYSTEMIC FAILURE

Over several decades, technical progress with enhanced electronic devices, high micro-processing speed and high storage capacity have unleashed a wide range of commercial products designed to attract sales and broaden the range of revenue-earning products available for the domestic consumer. Expectations have been fueled by what engineers are able to design, build, and deliver. And here we depart from physical laws and forces and the evolving development of ever-more complex machines to broach an area of engineering infrequently discussed: the pressure from commercial operators to engage engineers with satisfying not necessarily a market demand but a potentially marketable opportunity arising from the technology.

There is a tendency to add, attach, or build on to existing motor vehicles an increasing array of devices to attract a market niche or advantage over a competitor; it happens within the world of finance and it certainly occurs where engineers are required to provide increasingly sophisticated sales points for specific models or makes of vehicle, in order to attract the buyer and leverage a sale over competitors. This brings its own challenges to engineers as demands increase in an environment where competition is great and profit margins are small.

Fuel pump design

One classic example stands out: the humble fuel pump. Controlling the flow of a highly combustible fuel is a precision operation demanding reliable safety features and back-up devices. There are two primary components to the operation of a standard fuel pump: 1) the powertrain control module (PCM); and 2) the fuel pump relay. Direct control of the pump, which is a demand-driven component and has no autonomous operation, is through the relay providing or removing electrical power to operate it. The PCM, however, controls the pump by providing the relay with a ground. When the motor is running, the PCM turns the relay on, but if it does not see that the motor is running it will not provide a ground and there will be no power.

Overall, the PCM exists to monitor all signals from the engine compartment and is the primary controller for ensuring the smooth and even running of the engine. Individual manufacturers make their own selection as to which indicators signal effective operation of the engine to the PCM and some reference pulses come from the ignition module, which are used for ignition timing.

The engine control unit for a powertrain control module in a 1966 Chevrolet Beretta, electronically connecting the engine to the transmission control unit.

Other manufacturers might employ ***crankshaft sensors*** or ***cam sensors***, and these are used in different ways. The specific mapping of the sensors is key to determining which sensor suite a manufacturer selects. When a vehicle is standing motionless the fuel pressure gradually decreases and the engine will not run. When the ignition circuit is activated, there is a delay of a few seconds until the position in the ignition circuit is reached, which starts to crank the engine. Within less than a second, the engine fires and the vehicle is ready to roll.

Inertia switches

A lot of cars, increasingly so in large and less expensive popular models, have an ***inertia switch***, which is there to cut off the fuel supply in the event of an accident. The Ford company was well known for placing this in a specific position in all its models—beneath the passenger-side foot area—until it was noted that some passengers would accidentally kick that part of the floor, which would cause a reverberation and activate the inertia switch, shutting off the fuel supply by cutting power to the pump. This happened because the inertia switch was a circuit breaker and under hard impact from an accident shut off the supply and prevented leaking fuel spraying volatile fluid over a hot engine and causing a fire, or else spraying fuel over the road surface, impeding rescue services.

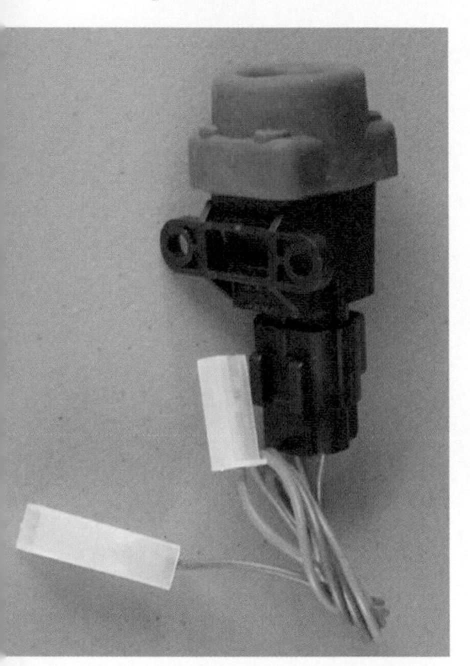

An inertia switch fitted to an automobile to cut power to the fuel pump in the event of a violent maneuver, registered by a g-force showing a shock-curve caused by a collision.

Although there are several mechanisms for controlling power to the fuel pump, in almost all motor vehicles it is the PCM that will decide when to do that. Some will vary the voltage to the pump so as to control the fuel pressure but increasingly there are now many sensors on a car that will respond to the shock spectra of side, front, or rear impacts.

Shock sensors are required by law in certain vehicles, and there is an increasing call upon the engineer to broaden the range of conditions under which they activate and, in effect, demobilize the vehicle. They usually operate on the basis of a proof mass, or small weight, contained within a spring-loaded cage. A shock in any axis attributed to a collision will cause a sudden movement of the mass relative to its position on the cage. This will spring open the cage itself, actuating an adjacent switch to cut the relay to the fuel pump. This provides a very simple but effective description of a low-scale systems integration. However, the spectra upon which the springs are calibrated must closely follow the shock curve—as well as the magnitude—of a shock caused by an accident and nothing less.

In one make of car available on the market at the time of publication, on occasions the inertia sensor could be triggered when the vehicle in question rumbled over a railway crossing, where the shock and vibration carry the same spectra as that from an accident, cutting power to the fuel pump and bringing the car to a stop with its rear end hanging over the track.

The majority of car drivers are unaware of the location of the sensor, which can easily be reset by activating a button on the top of the box.

Other applications for inertia switches

Inertia switches are also applied to air bags activated by an abrupt shock, but the spectra in this instance are set to a higher shock level. The most sensitive are inertia-reel seat belts, which lock during an abrupt braking motion, temporarily restraining the occupant in a firmly seated position. In this case, the offloading of deceleration unlocks the inertia reel and frees the seat belt.

Inertia switches are common in several operating systems at a hardware level, and are a classic example of how engineering can take standard commercial-off-the-shelf (COTS) products and mate them together within a matrix of bespoke engineered designs. Inertia switches are ***enablers of conditions*** (preservation of safety within a hazardous situation, such as a car crash) or ***enablers of activation*** (where systems are made active by inertia measurements). It is common for artillery shells and other ordnance to be fused (explosively activated) by an inertia switch responding to the acceleration caused when the round is fired, thus preventing the shell from being armed during routine handling or while being transported.

Inertia switches operate in reactive mode but can control proactive actions and introduce us to the complex subject of ***engineering control systems***, where connecting integrations are realigned to improve the operating potential of an engineered system. There are both ***open-loop control systems*** and ***closed-loop control systems***, but all have a common function: to manage, command, or control one or more devices by control loops. These can range from a simple thermostat on a home heating system to an integrated systems management role where a range of different machines or systems are managed in an integrated and co-ordinated manner.

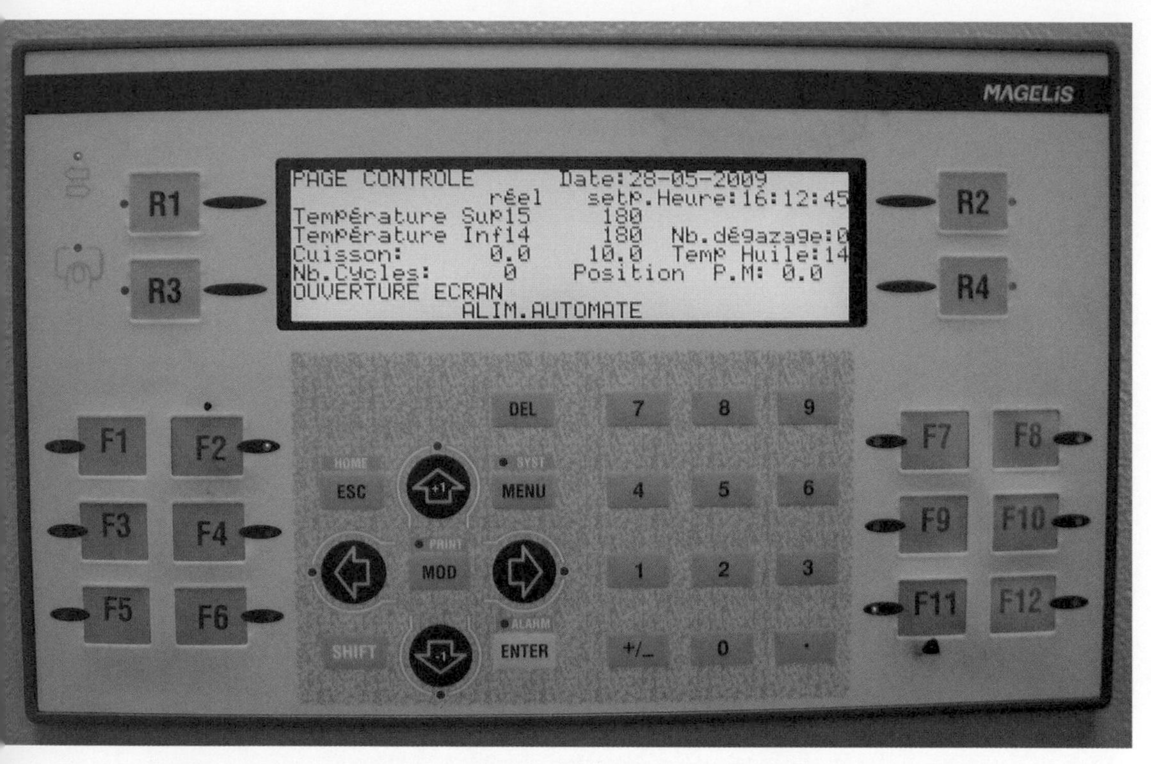

The control unit for a hydraulic heat press, with integrated software for function control and emergency cut-out switches in the event of a malfunction threatening the safety of the operation.

KEEPING CONTROL

Control loops are essential for the effective operation of machines, machine-tools, and larger integrated systems. The open-loop type is separate and independent of the process it controls, an example of which could be a timer switch for a domestic central heating system. In this, the variable process is the rate of temperature change in the void filled by the heating system, the timer operating the boiler for a fixed duration being entirely independent of the thermal setting. A closed-loop system depends upon the state of the processes measuring the variable. Here, the variable is the changing temperature of the environment served by the boiler and the enabling controller is a thermostat, which is unaware of the duration at which it will be set to operate—by the open-loop timer. Interconnected, they manage two sides of a common system.

Control systems require a ***feedback loop***, which regulates a variable around an adjustable ***set point*** (SP). The SP may be a temperature or a speed measurement, the latter in the case of cruise control in a motor vehicle, where a speed set by the driver is moderated by a ***proportional integrator*** (PI). Operating in a feedback loop to control other systems, the algorithm modulates the speed of the vehicle within pre-set parameters so that the engine is controlled not by the driver but by the PI. Clearly, only closed-loop control systems, and not open-loop systems that operate in pre-set ways, can be managed like this.

Hysteresis controllers

Another form of intervention is the ***hysteresis controller***, which operates only through on or off switching. This switch is typified by a bi-metallic strip that consists of two strips of different metals, which respond to heat by expanding at different rates. These could be steel and copper or steel and brass. The different expansion rates impose leverage on one, forcing the bi-metallic pair to bend in favor of the one with the greatest coefficient of thermal expansion. This action bends the strip as it expands at different rates and biases the strip with the greatest expansion to the outside of the curve, or to the inside as it cools. Bi-metallic strips or coils are used in a variety of devices, most notably thermostats, the coil changing the expansion of the metal into a circular movement due to the helicoidal shape it adopts in response to temperature changes.

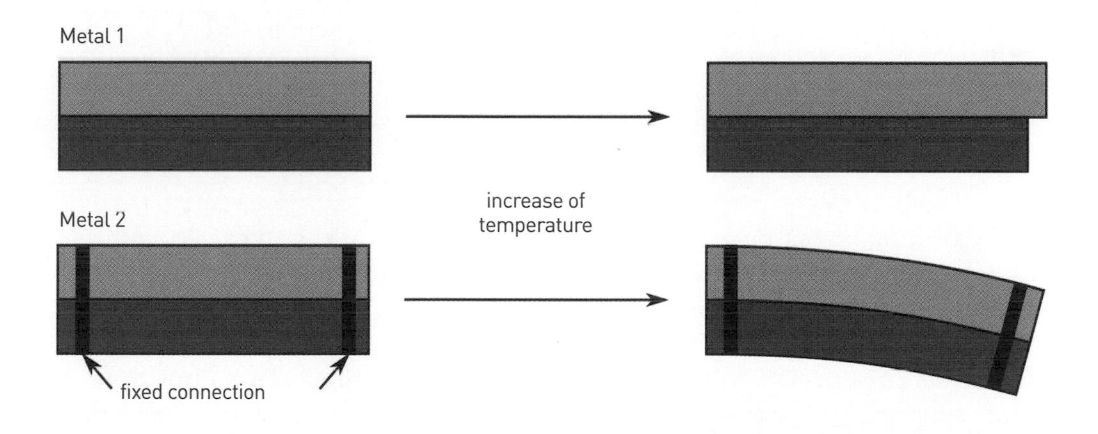

A bi-metallic strip showing two metals, each with a different coefficient of expansion that forces a contact should heat indicate the need to shut down the electrical operation of equipment.

Programmable logic controllers

Another form of systems management has its origin in ***ladder logic***, which was traditionally used as a means of documenting and sequencing a series of relays for repetitive manufacturing or logic routines. Today, much of that is covered by ***programmable logic controllers*** (PLCs) and benefits from developments in electronic and digital controls. PLCs are digital computers, hardened and ruggedized to manage repetitive but critical systems or groups of systems that lie at the core of several industries. These can include power stations, robotic storage and distribution warehouses, computerized manufacturing, and assembly lines. They depend upon reliability and low failure rates, providing repetitive but intermittently programmable control functions. They do this through the use of ***actuators***.

Actuators

Actuators are simply components of a machine designed to operate as a mechanical or electrical controlling mechanism that can be either a control signal or a source of energy in a fully mechanical function. The primary types operate through electric current, hydraulic fluid systems, or pneumatic devices. Electric actuators can come in a variety of forms, including:

- *Electromechanical actuators*—these power an electric motor to convert electrical energy into mechanical torque.
- *Electrohydraulic actuators*—an electric current provides torque to a hydraulic accumulator, which transmits an actuation force. Still embraced as an actuator, they can also be used to turn on or off small machines that operate electrical equipment or machines for performing tasks in the road-building industry, for instance.

Hydraulic actuators have a disadvantage in that liquids are almost always impossible to compress, and so these applications have a limited rate of acceleration despite being able to exert large forces. Hydraulic actuators invariably consist of a hollow cylinder and a piston connected to a spring, so that when pressure is relaxed the piston can return to its original position. A ***double-acting hydraulic actuator*** is one in which a fluid is supplied to either side of a piston, the movement being controlled by the force of the liquid in either direction, which is usually a linear path. These types of actuator are better suited to the movement of large masses utilizing strong forces and have few mechanical working parts, enhancing performance and increasing reliability.

Pneumatic actuators, on the other hand, have a wide range of applications and can apply considerable force from small changes in pressure at the point of activation. Compressed air or the use of a vacuum is usually employed at a high pressure when used with a diaphragm and for rotary or linear motion of a piston or drive arm. And because the power source need not be stored, response time is quick and instantly available. Working in conjunction with a valve system, a pneumatic actuator can considerably increase the applied force—at least two- or even threefold. Most lifting devices in relatively small engineering operations or workshops use pneumatic actuators because of this factor alone, and because the output pressure can be increased by raising the size of the piston. They are also a good selection should the air pressure available be relatively low. In theory, with an output of just 100 kPa, a weight of 992 lb (450 kg) could readily be lifted off the ground, though in practice the valve stem for a pressure scale of this size would be too small and would fracture.

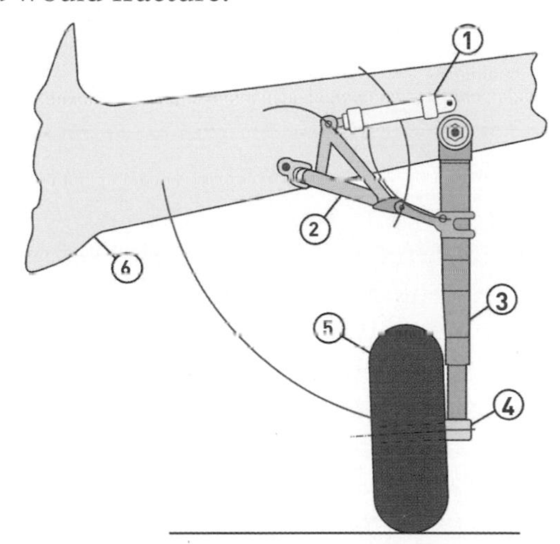

The hydraulic landing gear of an aircraft—schematic view of the retracting mechanism. Filled circles are fixed relative to the airframe. Colored arcs denote the locus of points. 1 hydraulic ram; 2 hinge mechanism; 3 strut; 4 wheel boss; 5 wheel; 6 fuselage/wing.

A pneumatic butterfly valve with moisture removed and a small quantity of oil introduced to prevent corrosion and to lubricate mechanical parts.

While there are some exotic forms of actuator, including ***electromagnetic*** and ***3D polymer printed systems***, they all face the same problem: engaging devices to convert circular motion to a linear drive. For this, ***lead screws*** produce a workable solution; but the use of ***screw jacks*** and ***ball screws*** are common and operate on the same principle as the basic screw, which leaves their stresses and tension levels well within known and measured parameters. This adheres to the first order of design engineering: always select an existing, proven, and reliable system for a new machine or collective set of machines. Sometimes the age-old empirical design of wheel-and-axle energy conversion is just as effective as a sophisticated solution. And that's another maxim of machine engineering: simplicity trumps complexity in terms of reliability and efficiency.

SYSTEMS ENGINEERING

As we have seen, engineering mechanical systems is all about machine integration bringing complex and innovative design to an operational level, often utilizing seasoned components and proven subsystems. It has essentially underpinned technological development throughout the 20th century. But halfway through that century, a new concept emerged that realized a different way of working—***systems engineering***. Its origins are argued over, but it really began in the early 1940s with the definition of a fast-track approach to producing advanced engineering systems in the briefest development cycle possible. And it led to some big errors, which 21st-century engineers should recognize for fear they may recur.

The realization of systems engineering in its purest form arose in the USA during World War II, which for the Americans was from the end of 1941 through to the signature of the Japanese surrender in September 1945. There was a need for rapid expansion in technological capabilities, and the old and certainly ponderous way of bringing new war machines into production was too slow for requirements; the US had not had the same urgency of preparation that was now required, so a compression of developmental steps was introduced. Systems engineering embraced a wider and more holistic analysis of what was required and presaged an era in which "systems design integration" became the product—a policy where separate elements of an entire system were developed in an integrated way.

Concurrency

When a new US fighter was developed on demand for a fresh requirement, separate parts of the aircraft were assigned to different design offices: the wings, fuselage, engine bay, tail section, undercarriage, crew compartment, and armament were all designed separately on instructions from the chief designer, who had drawn up the overall configuration. When this was seen as too unwieldy, an integrated approach began to be implemented in the 1940s, though it would take a further 20 years to be realized. What did result immediately from the dilemma of too few weapon systems and too little time was the emergence of the approach soon known as systems engineering, which rapidly took hold. In that concept, a concurrent set of steps was implemented to cut bureaucracy, eliminate intermediate stages, and cut wasteful steps between the origin of the concept and first operational trials.

This approach spawned what became known as ***concurrency***, where: (a) design, (b) development, (c) test and evaluation, (d) initial deployment, and (e) maximum production blended into a seamless continuity in which two or three of these steps were in parallel and not sequential, where previously completion of one stage resulted in a shift of effort to the next. To achieve this parallel sequence, it was vital to understand the critical path and to map that as an overlay on the development schedule. By doing that, it was possible to place elements of the engineering design on a critical path and flag them as potential showstoppers. However, where new and groundbreaking designs were involved, this method produced a concurrency policy that ran the risk of comprising the fluid flow of stages (c), (d), and (e).

A towering example of that was the Convair B-36, a very large post-war bomber which was so revolutionary that development overran and occupied much of the time the aircraft spent in operational deployment. The reason for this was that the project bridged two essential sequences in aeronautical engineering—the age of the reciprocating engine and that of the reaction engine—to such an extent that the B-36 started out with six piston engines and had four jet engines added in underwing pods halfway through operational deployment. In that regard, concurrency ran right through the entire history of the aircraft whereas, in theory, it should only have covered stages (a) through (c), above.

The product of a concurrency programme, the massive Convair B-36 saw service from the late 1940s to the late 1950s. With six "pusher" engines and propellers at the training edge of the wing, later it would receive four jet engines in two pods under the outer wing sections. Note the dull magnesium panels on various parts of the structure.

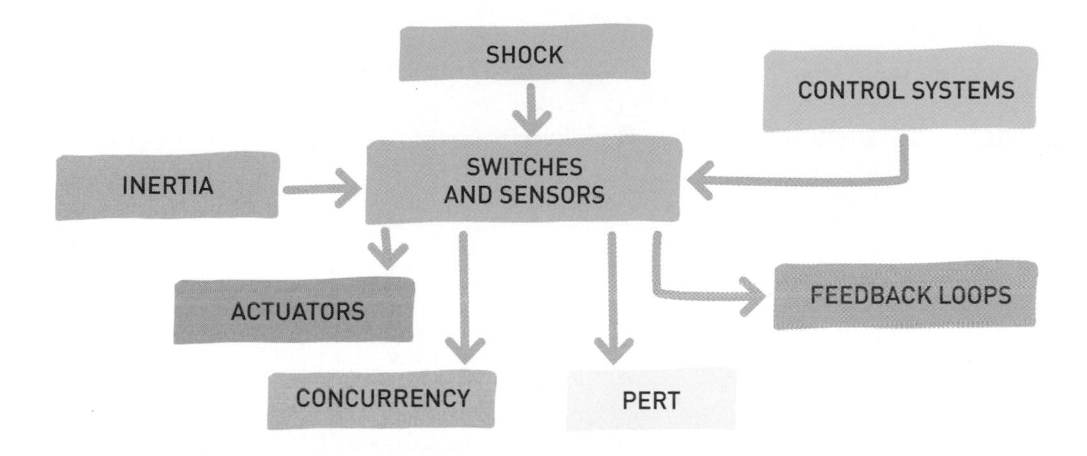

PROGRAMME EVALUATION AND REVIEW TECHNIQUE

By the mid-1950s, the systems engineering concept had been refined, and from experience during the war those practices were applied to the development of the first ballistic missiles under the Polaris programme, which also introduced a new management tool. This lay at the core of what has matured into the present system, and emerged via a program management tool known as ***program evaluation and review technique*** (PERT). PERT quickly became the benchmark for managing all the major aerospace and defence projects of the period.

PERT introduced a formal discipline into fixing a realistic end date and to back up from that to the start date, identifying key schedule limits for the intermediate steps—essentially, red lines that had to be met to achieve the end date. Key to this was to identify three things:

1. The shortest amount of time necessary to meet each intermediate task.
2. The most probable amount of time the schedule would take.
3. The longest time things might be expected to take should there be serious disruption to expectations.

In that regard, by backtracking from worst-case probabilities, a set of contingency routes were made available through pre-planning milestones. But there were consequences, and those will be discussed in the next chapter, in association with reliability and risk mitigation, both of which were vital for mechanical engineering.

Chapter Eight

WHAT IS RELIABILITY?

Assessing and Ensuring Reliability—Reusability and Reliability—Stepping into the Unknown—The Flaws in the System

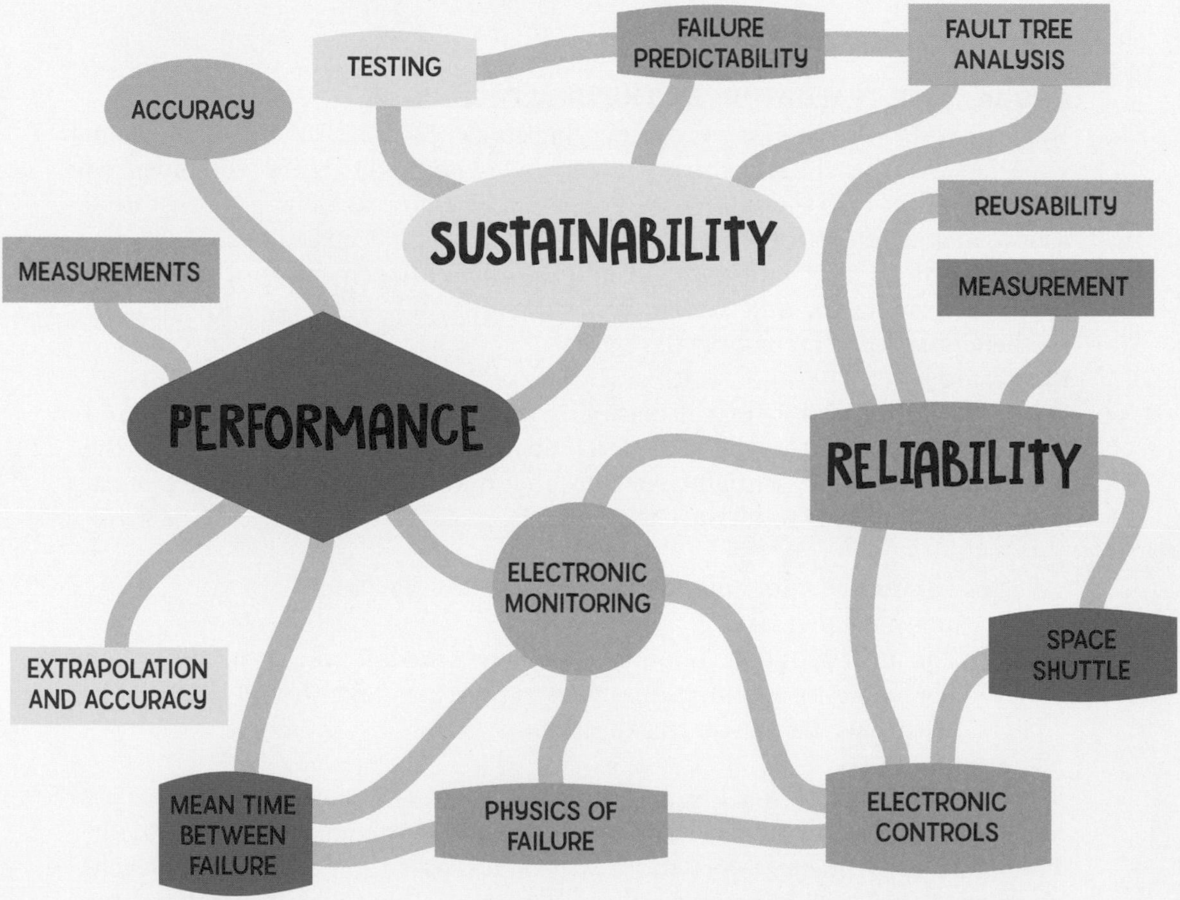

TESTING
FAILURE PREDICTABILITY
FAULT TREE ANALYSIS
ACCURACY
SUSTAINABILITY
REUSABILITY
MEASUREMENTS
MEASUREMENT
PERFORMANCE
RELIABILITY
ELECTRONIC MONITORING
SPACE SHUTTLE
EXTRAPOLATION AND ACCURACY
MEAN TIME BETWEEN FAILURE
PHYSICS OF FAILURE
ELECTRONIC CONTROLS

ASSESSING AND ENSURING RELIABILITY

One of the most important aspects of mechanical engineering is the balance between performance and reliability, which has been implicit within design and technology throughout the Industrial Age, and never more important than today. The classic definition of ***reliability*** is the ability of a system or a subsystem (even a component) to operate under the conditions rated for that element, and is a measurement closely associated with the probability of success at time T and denoted R(t). Separate sets of data scale the probability of reliability through several discrete parameters, including:

1. *Availability*—continuous readiness to conduct a defined task.
2. *Testability*—the degree to which the system can be measured.
3. *Maintainability*—the lowest frequency of failure and the greatest ease of support to sustain availability.
4. *Maintenance*—the ease under challenging circumstances of conducting routine servicing.

Sustainable reliability

Before we examine the way reliability is measured and the manner in which it can be both rated and calibrated, we must first examine one particular aspect. In recent decades, the phrase ***sustainable reliability*** has been heard, and this is key to the most critical condition for reliability. It means the ability of a system to accept some failures with discrete components or subsystems while still maintaining resilience in its primary operating objective. This is a robust form of operability where ***failure detection*** and ***fault tree analysis*** become crucial in determining the value factor in a given system or subsystem.

Fault tree analysis

Fault tree analysis (FTA) is a key in safety and reliability engineering, and employs a form of algebra known as ***Boolean logic*** in which variables, such as deduction of true or false, are denoted by 1 and 0 respectively. This form of analysis is critical in engineering where the consequences of failure are catastrophic on a colossal scale, perhaps involving several million casualties.

Types of failure and their analysis

A *two-dimensional failure* on a hypothetical model we could use might involve the linear consequence of an accident to a nuclear power station in which a very large number of lives could be lost from a single error, system failure, or collapse of an engineered subsystem. At a lower level of collateral consequence, the catastrophic failure of a major control system in an airliner in flight would be the same linear consequence, but here involving at most several hundred people, which, although tragic in itself, is not on the same scale as a massive leak of poisonous radiation.

However, the modern engineer is occasionally tasked with developing a system that operates in a ***three-dimensional*** setting, where a failure that in itself is not catastrophic can allow a consequential collapse of the very reason why the system was developed in the first place.

A typical example that we can use was the very origin itself of fault tree analysis, developed in the early 1960s at the Bell Laboratories for the US Air Force Ballistic Systems Division. It evolved as a requirement to ensure maximum reliability and safety with the launch control systems for the Minuteman missile. This was the strategic nuclear deterrent, and in particular the intercontinental ballistic missile (ICBM), a land-based strategic missile capable of sending a thermonuclear warhead across intercontinental distances. The development of the ICBM employed systems engineering to fast-track a capability that would have taken much longer using early concepts of project management.

The Air Force wanted a missile that could "launch on warning" of a pre-emptive attack by enemy ICBMs and their submarine-launched equivalents. Engineers

Now considered a fitting legacy of architect Eero Saarinen, the Bell Laboratories building in New Jersey conducted much of the fault tree analysis concept later adopted as standard throughout the engineering fraternity.

were therefore required to produce a missile capable of remaining in storage on a continuous basis over many years, yet be ready for launch within less than a minute. The fault tree analysis approach was used in the engineering design of the launch control facilities, where each underground station could launch ten missiles. The missile itself had been developed using the systems engineering process, and the underground launch control center was equipped with electronic equipment that met the same standard. The integrated system was subject to a fault tree analysis that demonstrated a new concept in engineering design for reliability known as ***fail-operational/fail-operational/fail-safe*** (FOFOFS). The FOFOFS concept employed sophisticated modeling techniques to which the engineer, in designing the hardware and integrated software, was required to build an interconnected and highly complex machine assembled through separate, albeit integrated, programs.

The aerospace manufacturers who developed and designed the Minuteman ICBM system operated to a higher specification than that which had existed in any defense program up to the beginning of the 1960s. The consequences of it not operating as required meant that it was invalidated as an essential element in the nuclear deterrent. Had tests and demonstration launches at a range in the Pacific Ocean displayed an inability to perform the publicized mission, the consequences for international tensions and, just possibly, the exacerbation of international tensions beyond diplomacy to war could have been catastrophic. This type of burden placed upon the engineer

The now aging Minuteman missile was the product of a systems engineering culture that significantly cut development time and enhanced reliability and performance, universally accepted throughout the defense industry.

had precedence in the nuclear industry (about which much more in the next chapter), but introduced absolutes in performance and reliability that literally rewrote the rule book on standards.

The FOFOFS concept was not known as such for more than a decade until another program brought the highest level of reliability involving a new and untried concept—the reusable Space Shuttle. This evolved at the end of the 1960s, was developed during the 1970s, and began flying in the early 1980s. The FOFOFS concept was required to ensure a largely fail-proof system, not based on the potential for its impact on the probability of global war, but on the safety of the crew in the most prestigious American flagship aerospace project since the Moon landings, which had taken place between 1969 and 1972. In the case of the Shuttle, this meant a complex and highly sophisticated reliability and safety analysis involving a unique approach to engineering safety assurance.

REUSABILITY AND RELIABILITY

The Shuttle program was a game-changer for engineering design and planning because it did not involve the application of a separate set of hardware and associated electronics for saving life in the event of a malfunction. It was a case study involving failure analysis and reliability because it introduced the integration of optimized mission performance and crew safety within the same systems and subsystems. Prior to this program, US manned space vehicles had been carried into orbit on rockets that initially had been developed as ballistic missiles and adapted as space launch vehicles. Only the Saturn rockets used in the Apollo program were built specifically to launch big payloads into space.

With those earlier adaptations from missiles, managers had at first believed that all the controls for launch and getting into orbit should be under the authority of the pilot, or the commander in a multi-person ship. But engineers argued that events happened too quickly for a single human to gather and co-ordinate all the information about the rocket's performance—only machines could gather information from several thousand data points and scan these within a fraction of a second. It would therefore be impossible for a human to make a proper decision about the flight trajectory and the vehicle's overall performance so as to maintain safety. It was decided to separate the powered-flight functions from the activities of the spacecraft, leaving the latter under the control of the on-board commander only after the rocket delivered it to orbit. That required a separate computer-controlled command system in the rocket itself for continuously extracting data from those several thousand data points and making a decision about whether an escape system should be activated.

Before the Shuttle, US manned space programs had crew escape systems carried external to the design of the spacecraft, on the basis that, if the prime hardware started to fail, a secondary set of hardware and equipment would take over and ensure the crew's safety. But the core function of the Shuttle was to demonstrate a level of reliability that would render unnecessary the crew escape equipment carried on previous spacecraft. These had involved escape rockets, either for wrenching the crew capsule free of an ascending rocket running amok, or ejection seats with which it was believed the crew could escape an impending inferno. In an unprecedented shift, adopting the practices of the aviation industry, Shuttle systems would be designed with the dual role of conducting an optimized mission and of being sufficiently flexible to operate as a means of preserving life—the true function of a "fail-operational/fail-operational/fail-safe" approach.

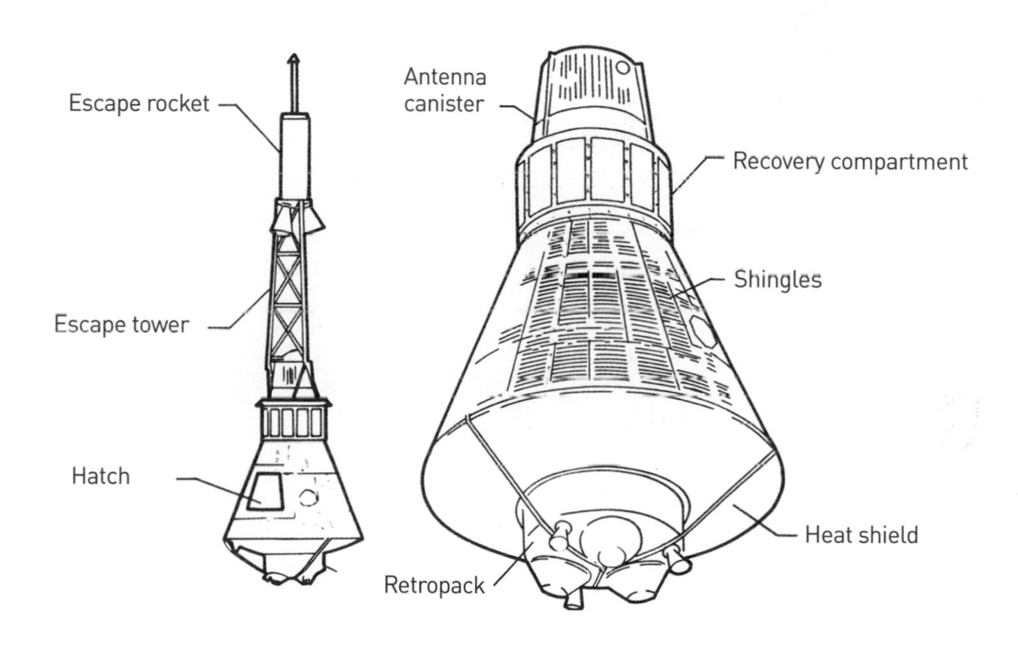

In early spacecraft built to carry humans, a launch escape tower was provided to carry the capsule free from a rocket that showed signs of catastrophic failure during its way up into orbit, the escape system itself being jettisoned when no longer required. This enhanced safety, by using a completely independent system to ensure the survival of the astronaut.

Consisting of a winged orbiter, two solid rocket boosters, and an external tank, the NASA Shuttle dispensed with a separate recovery system and instead adopted an in-flight abort concept whereby the winged orbiter would return the crew to a safe landing in the event of a systems malfunction—the fail-operational/fail-operational/fail-safe approach.

What this meant for the Shuttle was that the vehicle was designed to ride out two tiers of failure and still retain the ability to bring the crew back home. Aerospace engineering has always carried higher consequences for failure than most categories in which engineers are employed, but the pressure to design and build a system true to FOFOFS standards was revolutionary. It introduced new ways of defining safety and reliability, and drew considerable experience from the aeronautical industry. There had been several instances when secondary, safety-related, hardware had saved lives in the space program, notably some Russian flight aborts where a separate escape system had brought the crew safely back to the ground when their launch rocket had deviated and threatened catastrophe long before reaching orbit. Dispensing with that concept, the Shuttle system resulted in a loss of life when seven crew members died after the *Challenger* disaster in January 1986, the crew unable to escape an orbiter strapped to an exploding propellant tank.

The digital brains of the operation

Overall, the design of the Shuttle's systems configuration was dictated by this FOFOFS requirement, and the electronic brain of the vehicle was located in the five ***general-purpose computers*** (GPCs) carried in the obiter. Each GPC was composed of two separate units: a ***central processor unit*** (CPU) and an ***input/output processor*** (IOP). All five GPCs were IBM AP-101 computers. Each CPU and IOP contained a memory area for storing software and data. These memory areas were collectively referred to as the GPC's ***main memory***. The central processor controlled access to the GPC main memory for data storage and software execution and carried out instructions to control vehicle systems and manipulate data. In other words, the CPU was the "number cruncher" that computed and controlled computer functions. The main memory of each GPC was non-volatile, meaning that the software was retained when power was interrupted. The memory capacity of each CPU was 81,920 words, and the memory capacity of each IOP was 24,576 words; thus, the CPU and IOP constituted a total of 106,496 words.

From mid-1990, the new, upgraded, general-purpose computer AP-101S—from IBM—replaced the existing AP-101B aboard the Space Shuttle orbiters. These upgraded GPCs allowed NASA to incorporate more capabilities and apply more advanced computer technologies than were available when the orbiter was first designed. The new design began in January 1984, whereas the older GPC design had begun in January 1972. The upgraded computers provided two and a half times the existing memory capacity and up to three times the former processor speed with minimum impact on flight software. The upgraded GPCs were also half the size and approximately half the weight of the old GPCs, and they required less power to operate. They consisted of a central processor unit and an input/output processor in one avionics box instead of the two separate CPU and IOP avionics boxes of the old GPCs.

The upgraded GPCs could perform more than 1 million benchmark tests per second in comparison to the older GPCs' 400,000 operations per second. They also had a semiconductor memory of 256,000 32-bit words; the older GPCs had a core memory of up to 104,000 32-bit words. Moreover, the upgraded GPCs had volatile memory, but each GPC contained a battery pack to preserve the software when the GPC was powered off. The initial predicted reliability of the upgraded GPCs was 6,000 hours' mean time between failures, with a projected growth to 10,000 hours' mean time between failures. The mean time between failures for the older GPCs was 5,200 hours—more than five times better than the original reliability estimate of 1,000 hours.

STEPPING INTO THE UNKNOWN

Assessing the reliability of the Shuttle was a hybrid between the traditional method of summing the observed reliability levels and taking a synergized prediction based on tests. There was only so far that engineers could go in taking historic data upon which to base their assessments of reliability and probability of failure. The Shuttle was the first of its kind and therefore without precedent from which to extract historic data, and one of a kind without parallel against which to compare it.

Another challenge was that the fundamental design requirement was reusability and that drove an engineering compromise in that the systems it employed had to be capable of repeated flights—the goal was up to 100 missions for each orbiter. That, too, had an effect on reliability and on fault predictions. Previously, components had been tested individually and then connected via what engineers call a "breadboard," whereby equipment is electrically and sometimes mechanically connected but laid out on a two-dimensional jig for integrated testing. The issue is that equipment can sometimes give a different response when connected in series or parallel to that presented by a component or subsystem tested on its own.

We should also note here that a separate set of hardware and software for abort contingencies was not reserved exclusively for the rocket and the spacecraft it carried. During the Apollo program, a separate back-up computer with its own programs was carried by the Apollo mother ship and the Lunar Module that took two crew members down to the lunar surface. The back-up system carried information that was manually updated on board at each stage of the mission, capable of taking over and carrying the crew to safety should the primary system fail and leave them stranded. What's more, the hardware in these Apollo vehicles was the same for both primary and back-up functions. Nevertheless, it was an approach rejected for NASA's new reusable spacecraft, which, in effect, incorporated its own launch vehicle.

The Shuttle systems used for running an operational mission were also engaged for an abort and this took a step away from the traditional separation of rocket, spacecraft and abort systems to an integrated FOFOFS concept standard in the civil aviation industry. And there was high risk because of that. Even after the major computer upgrade in 1991, the primary flight system had a storage capacity of 1 megabyte and ran at a speed of 1.4 million instructions per second. While this was more memory and much faster computing speed than could be achieved with the original 1970s-era Shuttle flight computers, it doesn't compare to today's desktop computers. The GPCs included 24 input/output links that collected the signals from the Shuttle's myriad sensors and sent them to the GPCs. The computers plugged the readings from the sensors into elaborate mathematical algorithms to determine when to swivel the three main engines during launch, how much to move the control surfaces on the wings for landing, and which thrusters to fire in space to set up a rendezvous with the International Space Station, for example. That process was completed about 25 times per second.

The Shuttle's computer-driven flight control system was a first for a production spacecraft. The fly-by-wire design, tested on modified research aircraft, did not have any mechanical links from the pilot to the control surfaces and thrusters. Instead, the pilot moved the control stick in the cockpit and the computers transmitted signals to the control mechanisms to make them move. The Shuttle system was so dependent on computers that a fraction of a second without them could have been catastrophic during the critical parts of flight.

Located on the forward wall of the lower deck in the Shuttle's crew compartment, the avionics bay contained the GPC units and is here seen without the closeout panels, which would comprise one wall of the living quarters immediately below the flight deck.

THE FLAWS IN THE SYSTEM

To recap, the two-dimensional failure tree involves a catastrophic failure in a system, subsystem, or component resulting in a consequential loss of the machine to carry out its function—and in the case of our "airliner" model, to crash. This is an unambiguous, linear cause and effect. The three-dimensional failure can remain inert, not even requiring the machine to move, but in failing to carry out its task incur casualties on a colossal scale, as demonstrated by the example above. Throughout the last 50 years, the engineer has been tasked with an ever-changing set of requirements for test and failure probability analysis, not least because of the advent of advanced electronics, the impact of which we will explore in Chapter 10. Moreover, the increasingly advanced sophistication of complex machines such as the ICBM and the Shuttle has driven an ever-wider series of demands as the probability of failure increases with the multi-faceted integration of new and unique systems. However, there are established ways of predicting failure—not all of which are reliable. These standards require a measure by which to rate their effectiveness and that involves a potentially complex, yet basically simple, measure.

Mean time between failure

Mean time between failure (MTBF) is one of the most important measures, by predictability, that can influence not only the performance of the system in question but also in contractor selection, increasingly an important aspect of what the modern mechanical engineer has to do. MTBF is defined as the time between inherent failures in either mechanical or electrical systems. It is possible to define this as an arithmetical mean between failures of a repairable system as well as a component. The measurement is usually made by running a series of identical components until they each fail, the MTBF being the mean of the elapsed time when each fails, that value being expressed as the ***mean time to failure*** (MTFF). That number is used as the quotable MTBF value and engineers will be required to demonstrate the longest average they can achieve. Based on an actuarial spread, the greater the number of component tests done in this way, the more accurately the number reflects real-world situations.

In a repairable, or self-repairable, system, however, the MTBF is judged to be the sum of the difference between the "down" times and the "up" times, when it is in-spec and out-of-spec, respectively. In this regard, the total declared operating time is the time the system was actually operating to specification and not the total elapsed time of the test running to the point where the component, or system, fails.

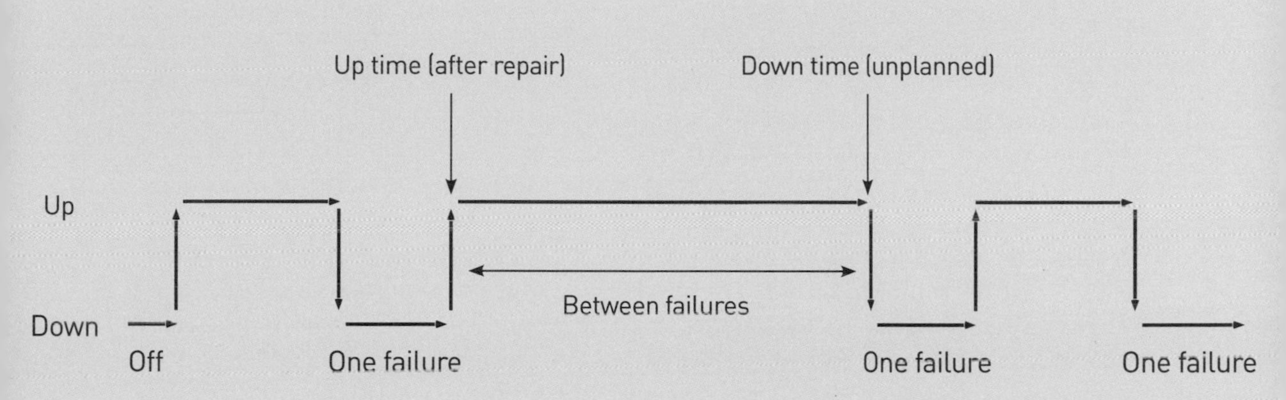

As a measurement of mean time between failures (MTBF), the value is expressed as the mean time of the average of three failures, known as the "up" time. The repair time can be calculated for autonomous self-repair or for manual repair by an intervening individual.

MTBF factors differ between mechanical and electronic systems, subsystems, and components and whether the subsystems are arranged in sequence or in parallel, which means that the qualifying criteria must be declared before the rating of different configurations of equipment can be assessed.

But there are potential flaws in the way reliability values, MTBF, and risk assessment are applied to unique machines and in particular to an arrangement of independent systems that, although perhaps having a reliable and well-logged history, operate in a completely different way when reconfigured. Reverting to our discussion of the Shuttle, predictions on reliability and risk potential within the total operating system fell far short of the trueness (mentioned in the previous chapter). When developed as an integrated system, the separate elements of orbiter, external tank, main engines, and solid rocket boosters were subject to conditions that had been simulated before first flight in a range of test equipment and through simulated environments. The design goal was to have a total Shuttle reliability of 99.999 per cent, implying a single failure of 1 in 10,000. This was far less than the failure acceptable in any other form of transport, but considered good for the technology involved. By way of comparison, the airline industry records only one crash in 4.4 million flights.

Readers are encouraged to research the many Shuttle technical issues that occurred during operational flights, on many occasions bringing the system close to catastrophic failure. Suffice to say here that with the loss of *Challenger* after 24 successful flights and the loss of *Columbia* after 112 launches, a considerable body of evidence was assembled to retrospectively revisit the failure, which was probably one of calculation based on an increasing amount of data available as the program progressed. At the end of the program, after 135 launches, the failure probability on the first launch was re-evaluated. Instead of the 0.001 per cent failure probability claimed at first launch, based on 30 years of experience with operating the Shuttle, the risk of failure was re-assessed at 10 per cent. This shocking statistic was a sobering lesson for the post-Shuttle space program, and is an object lesson in assumptions tinged with overconfidence that every engineer should be aware of in any application of electromechanical engineering to the real world.

There are many tools available to engineers in the 21st century that were not even dreamed about at the turn of the millennium. These include computational models requiring very large data storage facilities such as ***built-in-self-test*** (BITE), whereby machines can diagnose themselves at frequent intervals to conduct real-time analysis, ***failure-mode-and-effects-analysis*** (FMEA), the fault tree analysis we have already described, and the ***finite-element method*** (FEM) using the Monte Carlo method. The Monte Carlo method is essentially a range of computational algorithms that conduct random sampling to eliminate bias which could otherwise prejudice a prediction.

Later, we will look at the way in which machines can be made to analyze themselves, carry out a statistically based risk assessment, develop a predictable set of failure timelines, and carry out repair and risk-mitigation steps. This is on the way to discussing machine automation, rapidly becoming a central part of electrical and mechanical engineering and the basis for artificial intelligence (AI). None of these would be possible without the advanced levels of reliability and risk analysis that on a daily basis preoccupy the engineer in designing and developing new ways to provide infrastructure for a modern society. Furthermore, over time, engineers are tasked with satisfying the whims of an increasingly consumer-based culture with growing expectations and higher demand levels for retail goods. But it is at the core level of utilities—power and sustainable energy requirements—that the technological edge is being pushed further forward.

The sigma scale

Before we leave the subject, there is a category of engineering known as the ***physics of failure*** (PoF) which is worth study in its own right. It applies mainly to electrical and electronic engineering, as we will explore further in Chapter 10, but one significant element in it is the measure of probabilistic design known as the ***sigma scale***. This is a statistical measure of dispersion in data and uses the data history to make predictions about the probability of successful performance in the future. With a normal probability distribution, the number of defects predicted to occur will depend upon what is referred to as the ***sigma number***.

Sigma is the Greek letter for accuracy and the sigma number indicating the standard deviation begins with 1 and multiplies up to 6. Starting with sigma 1, it includes all data points between, say, 10 to +/-2 (in other words between 8 and 12). A 2 sigma measure includes 10 +/-2(2) (that is, all the data between 6 and 14). In other words, it provides an increasing probability of factoring in random as well as reliable data, such that a sigma 1 level show 69 per cent defects and a sigma 6 level shows 0.0000019 per cent defects. The higher the sigma number, the more defined are the known defects in the data. In this way, the sigma number does not reflect the reliability of a system from which data are extracted, but rather indicates the level of certainty in the "trueness" of the data themselves. The actual reliability predictor comes from the accuracy of the data used to obtain the real-world figure.

Chapter Nine

NUCLEAR ENGINEERING

Atomic Potential—Fission and Fusion—Realizing the Potential—Modern Nuclear Capability—"Clean" Energy—Types of Reactor—Nuclear Propulsion

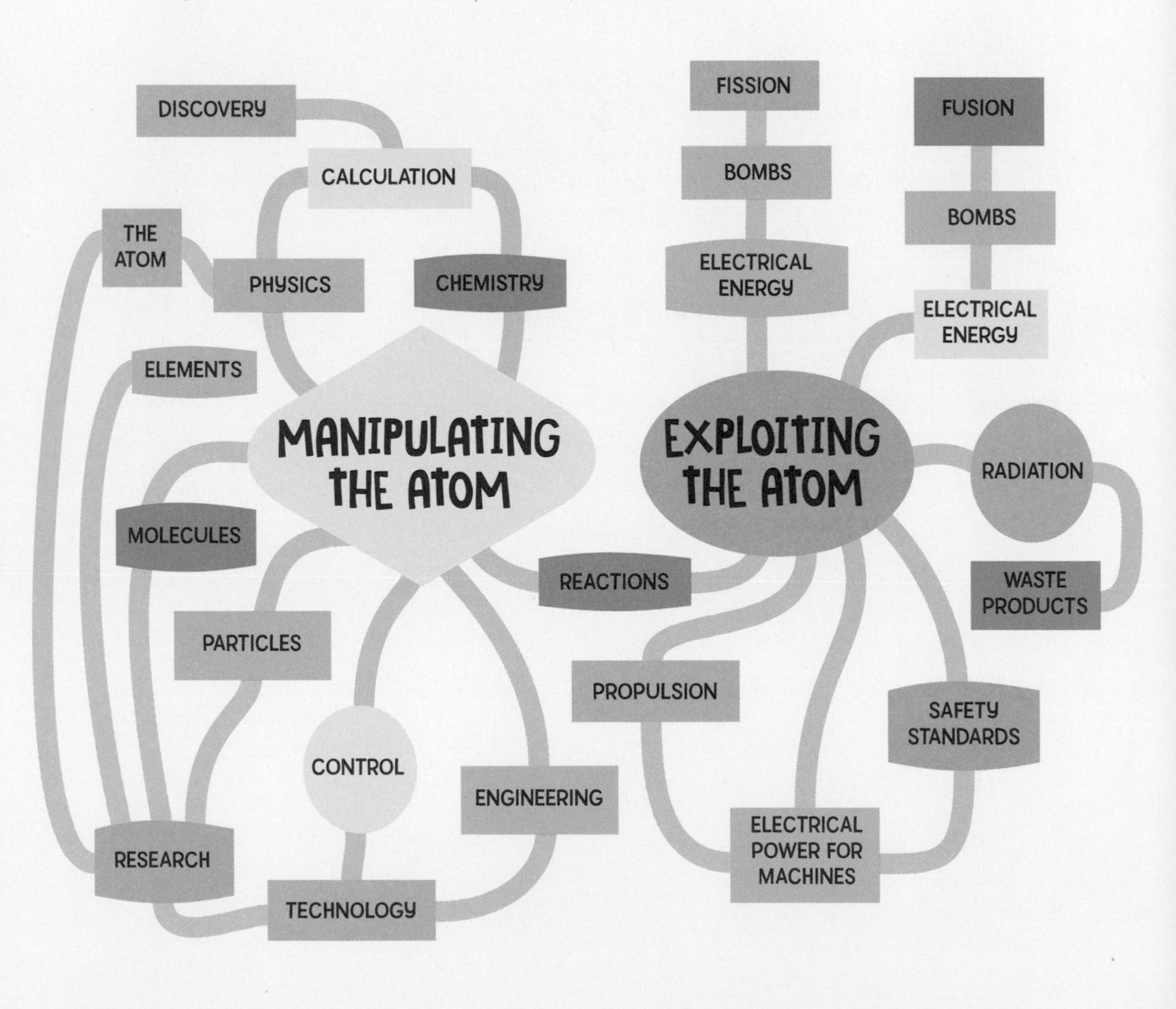

ATOMIC POTENTIAL

There is one field of engineering that touches several applications, ranging from the design and development of civil and commercial commissions to the cutting edge of military requirements. It embraces the field of utilities on one hand and the development and deployment of the most destructive instruments of war known to humans on the other. ***Nuclear engineering*** covers many separate fields, but all are catalyzed by one enabling scientific process: energy contained within an ***atom***.

Unraveling the physics of the atom occupied scientists for the first several decades of the 20th century until Otto Hahn (1879–1958) and Fritz Strassman (1902–80) became the first to split the atom (1938)—a ***fission*** that unlocked virtually limitless amounts of energy for engineers to harness and governments to invest in—primarily for electrical power and for an explosive device colloquially known as the ***atomic bomb***.

Today, ***nuclear energy*** is the bedrock of power production in many countries, presently providing 11 per cent of the world's electricity (424 GW) from 430 nuclear reactors, of which about one-quarter are in the USA, a country that gets 20 per cent of its energy from this method. At the other end of the spectrum, ***nuclear weapons*** remain a mighty force and a potent threat, but while they are considered a "deterrent," they will continue to be a part of the defense infrastructure in several countries around the world. Nuclear engineers embrace these very different applications and face some of the most exacting challenges in performance, safety, reliability, and robust risk mitigation because the consequences of getting it wrong are potentially catastrophic.

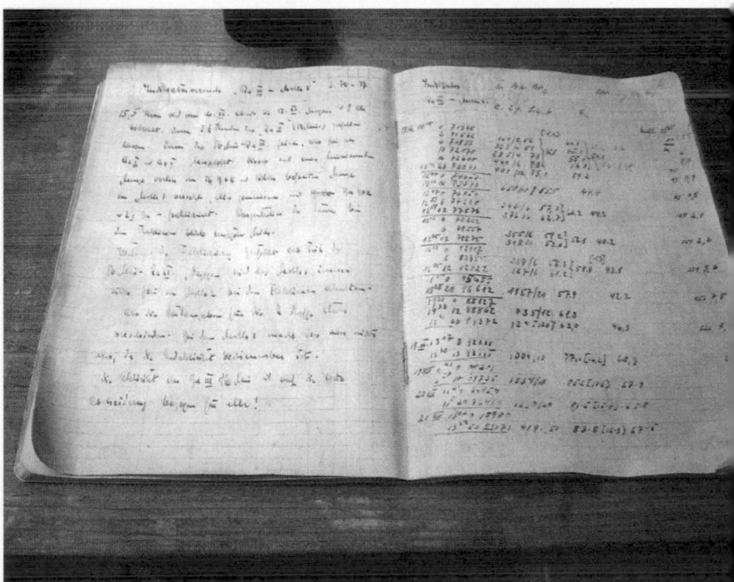

Otto Hahn's notebook of the experimental apparatus where he led his team, including Fritz Strassman, to discover nuclear fission in 1938.

In addition, there are several subsets of nuclear engineering, including the design and development of nuclear engines for naval vessels, research into nuclear rockets, and the development of highly advanced propulsion for uncrewed interstellar starships, which several space scientists expect to be the next major challenge for the exploration of space. Down on Earth, other aspects include nuclear medicine and medical physics, the handling and processing of nuclear materials, development of radiological and radiation protection devices as well as the production of small thermoelectric nuclear devices for providing electrical power in remote locations. All these diverse applications are possible through decades of research into the physics of the atom and the way in which science has played a pivotal role in producing the basic understanding of the atom so that it can be used for these diverse and challenging applications.

The core of CROCUS, a small nuclear research reactor at the École Polytechnique Fédérale de Lausanne in Lausanne, Switzerland.

As an ultimate application of nuclear propulsion, the bizarre concept of using detonating atomic bombs to propel a spacecraft to the far reaches of the solar system was an exercise in extravagant engineering, requiring technology that did not exist when proposed in the 1960s. This artist's concept shows the basic design as envisaged. Challenges posed by exotic and sophisticated concepts are never wasted, and push ingenuity and inventiveness to the limit.

FISSION AND FUSION

Nuclear engineering is about manipulating and exploiting the characteristic features of the atom and of subatomic particles. The fundamental understanding of matter began long before Hahn and Strassman first split the atom. Scientists had known for some considerable time that all matter is made up of atoms, of which there are more than 100 different types, each one defining a specific ***element*** such as carbon, iron, zinc, platinum, and uranium, to name but a few.

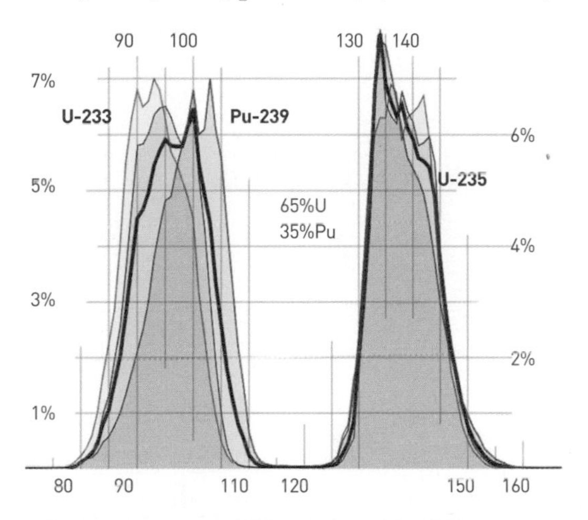

Yields by mass for thermal fission of uranium-235 and plutonium-239, which displays the energy yield of these two atoms. Fission splits the nuclei of heavy atoms to liberate energy while fusion takes the lightest atoms and liberates energy by bringing them together.

What is an element?
Elements are the fundamental building blocks of chemistry and of chemical substances that consist of individual atoms.

What is an atom?
The atom contains a nucleus of positively charged protons and neutrons which possess no charge, surrounded by orbiting electrons, each one of which balances the opposing charge of the similar number of protons. Individual atoms can be made to join together to form compounds, the smallest of which are molecules.

Each element has a characteristic number of protons in the nucleus, balanced by a similar number of electrons, and this determines the nature of the element. In chemical reactions, the neutrons have little effect but in nuclear reactions they are very important. Hydrogen has no neutrons, only one proton balanced by one electron, and is therefore the lightest element. However, a percentage of natural hydrogen can possess one neutron (deuterium) or two neutrons (tritium). Deuterium can combine with oxygen to make ***heavy water***, so called because the mass of the hydrogen isotope has a greater mass than natural hydrogen.

Mass defect and binding energy

Two types of nuclear reaction are possible:

1. Fission—the splitting of a heavy nucleus into a pair of lighter nuclei.
2. Fusion—the combination of two very light nuclei to form a heavier one.

The underlying principle in both cases is that the average net energy attraction between the nucleons (protons and neutrons) is smaller in the initial nucleus than it is in the products of the reaction. Determination of the ***mass defect*** (M.D.) can be calculated if A is the ***mass number*** and Z is the ***atomic number***, so that the nucleus contains Z protons and A - Z neutrons. If m_p is the mass of a proton, m_n is the mass of a neutron, and M is the actual mass of the nucleus, then the mass defect of a particular isotope is defined by: M.D. = $[Zm_p + (A - Z)m_n] - M$. The mass defect, which can be taken as the decrease of mass, which would result if Z protons and A - Z neutrons are combined to form a given nucleus, is a measure of the binding energy (B.E.) of that nucleus.

Atomic mass units

If the various masses are expressed in ***atomic mass units*** (amu), where mp is 1.00813 and m_n is 1.00897, multiplication of M.D. by 931 provides the B.E. in millions of ***electron volts*** (Mev). This factor is based on ***Einstein's mass energy equation*** using the appropriate units of mass (amu) and energy (MeV). Consequently: B.E. = $931\ \{[Zm_p + (A - Z)m_n] - M\}$ Mev. This provides the value of the energy released in the formation of a nucleus by a combination of the appropriate protons and neutrons, or alternatively, the energy that would be necessary to break up the nucleus into its constituent protons and neutrons.

Radioactive isotopes

In the original uranium-235, the binding energy per nucleon is about 7.5 Mev, which is the total amount of energy required to break up the uranium-235 nucleus into its 235 constituent nucleons. On subtraction of the higher value to the lower: uranium^{-235} → fission products + (235 x 0.9) Mev, showing that the fission of U-235 is accompanied by the release of 235 x 0.9 Mev, which is about 202.5 Mev of energy.

Because the fission process itself is accompanied by the liberation of neutrons, a chain reaction takes place to ensure sustained reaction. Only three isotopes are practical for the fusion process: U-233 (bred from thorium-232), U-235, and Pu-239. These substances are radioactive and have relatively long half-lives, which means they are moderately stable and will undergo fission by capture of neutrons of all energies, either fast (high energy) or slow (low energy). Just 0.7 per cent of all uranium is of isotope 235 and this, as well as plutonium-239, was the only nucleon used in early nuclear weapons.

Uranium sources

With an average concentration of about three parts per million, uranium is 40 times more abundant than silver in the Earth's crust, with total deposits estimated at about 110,000 billion tonnes, of which 58,000 tonnes are mined each year. More than a quarter of this total comes from Australia.

Uranium 238 accounts for 99.3 per cent of the naturally occurring isotope and has the longest half-life of all, at 4.468 billion years. But this isotope requires neutrons of at least 1 Mev to cause fission. While most neutrons produced in fission have higher energies, they rapidly lose energy in collisions, and sustained fission reaction is impossible. However, some fission does occur and the energy released can contribute appreciably to the total energy produced in nuclear weapons. Taking U-235 as an example, the fission cycle can be expressed as: U-235 + n → fission fragment + 2.51 neutrons + energy.

The number of neutrons produced in the fission process varies with the different ways in which the nuclei are split. The energy of fission neutrons ranges from quite small values up to 14 Mev or more, the majority having energies of 1–2 Mev. The average number of fission neutrons produced varies, too, according to the initiator used, from 2.51 per fission for U-235, to 2.60 per fission for U-233, and 2.96 per fission for Pu-239.

The fraction of neutrons escaping can be decreased or increased by increasing the mass. Because neutrons are produced, by fission, throughout the whole system but only lost from the exterior surface, the probability of escape will decrease as the volume/area ratio of the system is increased and for a given geometry this can be achieved by increasing the physical dimensions. The ***critical mass*** for a ***chain reaction*** is dependent on the nature of the fissile material, its shape, and several other factors. If v is the average number of neutrons produced in each act of fission and ι is the average number of neutrons lost by escape or other means, then v - ι is the number available to sustain fission, which in the following can be represented by k, so that $k = v - \iota$. For every neutron causing fission in one generation, k neutrons will be available to cause fission in the next generation. Hence, k will be less than unity ($k < 1$) in a subcritical system but will be unity ($k = 1$) in a critical mass and greater than unity ($k > 1$) in a supercritical system.

REALIZING THE POTENTIAL

Making functioning machines using these calculations and the ensuing liberation of energy, defined by Albert Einstein (1879–1955) in $E = mc^2$, clearly involved fission in the creation of a weapon of mass destruction and as a powerful means of producing electricity. The British played a seminal role in developing the theoretical base for each application and, before World War II, its government secretly debated whether or not to place major funding into an atomic weapon, for offensive purposes, or a comprehensive early-warning system using radar, for defense against potential enemy air attack. Requiring considerably less financial investment and of more immediate and practical application, radar won out and, while various agreements were raised with the Canadians for access to ***radiogenic materials*** for long-term plans embracing development of both a bomb *and* a nuclear power supply, any commitment from the British government would have to wait.

The long and tortuous path taken by the British to persuade the Americans to use their considerable financial assets and manpower to develop a bomb resulted eventually in the Manhattan project to develop such a device and to use British scientists and engineers on the team. An agreement forged between British Prime Minister Winston Churchill and US President Franklin D. Roosevelt resulted in US pre-eminence in the development of an atomic bomb while the British would have the post-war lead on atomic power for domestic purposes. That story is for another place but the significant role played by British scientists and engineers put the UK in a strong position after the war to develop nuclear energy—which it seized with enthusiasm as a world leader in the field.

The atomic bomb

The first atomic bomb was detonated under test conditions at Alamogordo, New Mexico, on July 16, 1945. It was followed by the dropping of the first

Nuclear fusion

Nuclear fusion is the opposite of fission: light nuclei are made to fuse by being brought together at extremely high temperatures of 50 million–500 million °F (10 million–100 million °C) to form heavier elements—a process that liberates great quantities of energy.

There are several different reaction chains possible but the most common are those between deuterium and tritium (D + T), because this reaction proceeds more rapidly at temperatures likely to be obtained by a trigger, that being the detonation of a fission device to start the process. An alternative fusion chain is between deuterium and deuterium (D + D) but the D + T reaction is 100 times more probable in the temperature range of 100–100 KeV, where 1 KeV equals 52.9 million °F (11.6 million °C). Moreover, a D + T reaction can be achieved with a lower reaction temperature than for other fuels. The principle reaction is: $D + T \rightarrow He^4$ (3.52 MeV) + n (14.07 MeV). For comparison, a tritium to tritium fusion provides less energy: $T + T \rightarrow He4 + 2n + 11.4$ MeV.

atom bomb, a 15 KT uranium device, on the Japanese city of Hiroshima on August 6, 1945, followed by a 23 KT plutonium bomb, of the type used at Alamogordo, dropped on Nagasaki three days later. As mentioned above, the critical factor for a chain reaction is determined by its specific mass and its geometry. Each device had a unique design and geometry for detonation, the Hiroshima bomb being of the gun type in which 141 lb (64 kg) of enriched uranium was brought to criticality by one hemisphere being driven into the other. The Nagasaki bomb had 13.6 lb (6.19 kg) of plutonium brought to criticality by implosion from 32 charges detonated in symmetry. The sphere was chosen because it has the greatest volume for the least surface area, and is a shape that will minimize to the lowest theoretical value the number of neutrons escaping unproductively. A critical mass of U-235 is 115 lb (52 kg), or 35 lb (16 kg) for U-233 or 22 lb (10 kg) for Pu-239. It is also possible to lower the critical mass if the density is increased, so that for a fissile material of radius R and a uniform density of p: $(p \cdot R)_{critical}$ = constant. For a fixed core mass, M, uniformly compressed, the density is given by: $p = 3M/4\pi R^3$. As a consequence, the critical mass is proportional to the reciprocal of the square of the density, as in: $Mcritical = k/p^2$.

Paradoxically, the energy released per atom is less for fusion than it is for fission. However, because the nuclei are lighter, the amount of energy per unit mass is three to four times that obtained from the heavier nuclei in fission bombs. Nevertheless, fission bombs are of relatively low yield and are the simplest form of device. Most of the early weapons were of this type but a combination of fission and fusion reactions came to populate most of the assembled arsenals.

Testing the atom bomb
The first full-scale test of a thermonuclear device was conducted under the name Operation *Ivy Mike* on November 1, 1952 on an island in the Eniwetok Atoll in the Pacific Ocean. A total of 11,650 people were involved with the test. *Ivy Mike* delivered a yield of 10.4 MT, causing large amounts of radioactive fallout. The fireball had a diameter of up to 4.1 miles (6.6 km), which was reached in seconds after the detonation. Amid shards of lightning created by the atmospheric effect, the mushroom cloud rose to an altitude of 10.6 miles (17 km) in 90 seconds and eventually flattened out at a height of 25.5 miles (41 km) spreading to a diameter of 100 miles (160 km) on a stem 20 miles (32 km) wide.

The Castle Romeo nuclear bomb test of March 27, 1954 produced a yield of 11 MT (equivalent to 11 million tonnes of TNT)—more than 700 times the yield of the bomb dropped on Hiroshima and a product of the manipulation of atoms following millennia in which metallurgists manipulated molecules.

MODERN NUCLEAR CAPABILITY

The story of the evolution of nuclear weapons from the fission bombs of World War II and the unimaginably destructive thermonuclear fusion weapons in today's arsenals is beyond the scope of this book. Suffice it to say that while the total number of nuclear weapons has declined since the intense days of the Cold War, there are still roughly 10,000 warheads and bombs ready for instant use by eight countries (Russia, USA, France, China, UK, Pakistan, India, North Korea, and Israel), with at least two more working very hard to get them. These weapons are more accurate today than they have ever been and several countries, including China, USA, UK, and France, already have funded programmes for further development and/or expansion of their nuclear deterrent forces.

Today, all forms of nuclear testing by detonation are banned but there are some indications that members of the "nuclear club" have conducted clandestine tests of very low-yield weapons—North Korea is an example. Nuclear engineers have developed sophisticated tools for monitoring such illegal activities and those include seismic and air-sampling instruments in addition to satellites in space designed to monitor potential irregularities. Most nuclear club members have signed an agreement banning tests and there is a robust verification procedure as both the USA and Russia have worked to dismantle weapons from their inventories restricted by arms limitation treaties. This is work for highly specialized nuclear engineers, as enriched uranium can be converted into low-grade fuel for use in power stations. Ironically, the two Cold War protagonists have exchanged these low-grade materials as trade, the USA buying from Russia such material for use in nuclear reactors. Under what was dubbed the "Megatons to Megawatts" program, fuel reprocessed from 20,000 nuclear weapons has helped to keep the lights on.

"CLEAN" ENERGY

Nuclear scientists and engineers carry out some of the most demanding work in the field of nuclear energy, due largely to the demand for "clean" energy. However, while some are engaged in working on the next generation of power stations, others are researching the possibility of nuclear fusion for totally waste-free power. If achieved, this would be the biggest and most environmentally friendly source of energy humans have ever created. Many would say it is long overdue. Yet the science is difficult and nobody has yet shown how such a fusion station could be built—which does nothing to halt the work of those who try. For the time being—and probably for several decades to come—the only type of nuclear power station is one based on fissile materials to develop the thermal energy to drive turbines and produce electricity for distribution through the grid. It employs the fundamental laws of thermodynamics, in that it converts one form of energy into another, and the entropy of a closed system can never decrease, operating on the principle that heat will flow from a high temperature to a low temperature zone—the latter being the point at which electricity is generated.

In purest terms, thermal power plants embrace both fuel burners and those generating power by a prime mover but here we have grouped nuclear power stations as embracing nuclear engineering, favoring the source of the energy rather than the end-point application. Other forms of non-nuclear power generation will be dealt with in Chapter 11.

Today, France leads the world in getting almost 71 per cent of its electrical energy from nuclear power stations while the UK gets less than 16 per cent and Germany plans to phase out all its nuclear stations by 2022. With the world's highest use of fossil fuel, China gets only 5 per cent of its electricity from nuclear power while Ukraine ranks high on the world list with 54 per cent—a legacy from the Soviet era. We should note, but only as a side issue, that a lapse in commercial and financial investment, as well as a political decision, shifted the premier role in power station design from the UK to other countries such as the USA, France, and China.

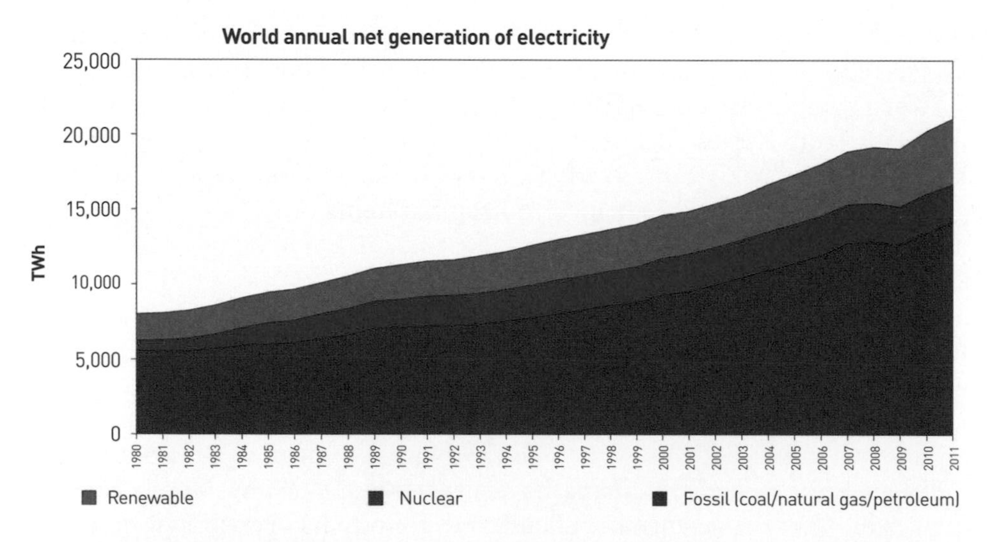

A challenge for engineers of the future, the inexorable rise of fossil fuel power stations and the consistency of nuclear power compared with renewables, which are less reliable but have greater environmental credentials.

A diverse role
Nuclear engineers embrace physics (for the basic operating principles), hydraulics (for water cooling systems), health and safety, processing, project management, quality engineering, and reactor operators. Like those of other thermal power systems, nuclear engineers are required to have degrees in either engineering or science. Specializations are legion and require a broad base of skills and professional experience. Employment opportunities are arguably unique to the field of electrical and mechanical engineering as it includes areas only peripherally associated with power production itself.

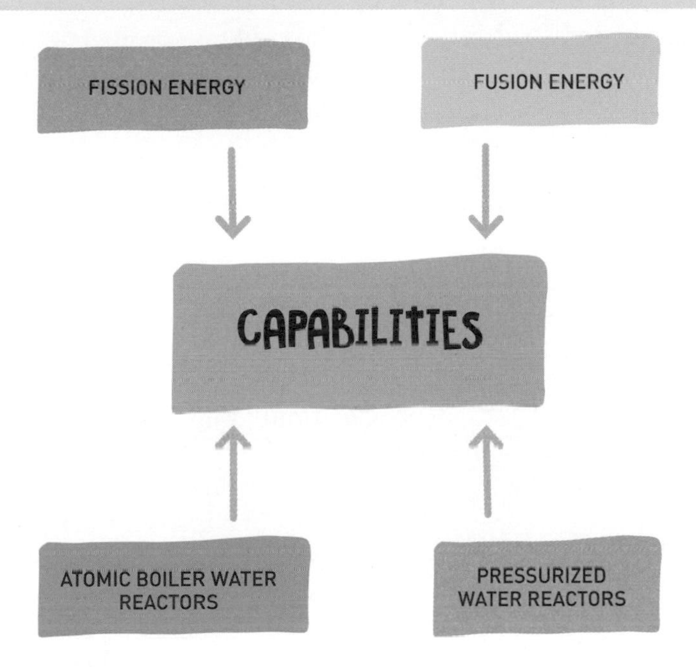

TYPES OF REACTOR

There are several different types of reactor, and legislative constraints and controls vary widely from country to country, while national preferences have slanted commissions of type across two basic categories—***boiler water reactors*** (BWR) and ***pressurized water reactors*** (PWR). The difference between the two is that a BWR uses the reactor core to heat water for turning into steam, which then drives the turbine, whereas the PWR heats water, which, instead of boiling, exchanges that thermal energy with a lower-pressure system, which is then turned to steam to drive the turbine.

There is another class known as the ***breeder reactor*** which was favored in the late 1950s. In this system, the reactor produces more fissile material than it consumes—made possible because the economy of neutron use is sufficiently high to deliver an excess. These have since fallen into disfavor because better and more efficient methods of uranium enrichment have been developed. Fast breeder reactors utilize uranium-238 while thermal breeder reactors use thorium-232 and have a thermal neutron spectrum using slow or thermal neutrons. Development with breeder reactors continues apace, however, and nuclear engineers are working on successive generations, which promise lower costs.

The pros and cons of nuclear energy

With such an emotive subject, claims and counter-claims by advocates and opponents of nuclear power generation muddy the facts for the casual observer. The truth is that while all fission power stations do produce waste, 97 per cent of this is considered such low risk that it is easily disposed of. The 3 per cent that is rated as toxic is stored in what are considered to be safe zones in storage facilities or underground. Low-grade radioactive waste decays to 1,000th the radiation level within 40 years. High-grade waste takes several thousand years to become harmless. Irresponsible claims to the contrary do little to raise the level of the debate but against this should be rated the industrial waste from cadmium and mercury, which are common in a wide range of appliances, and retain their toxicity forever.

The waste argument aside, there is no doubt that nuclear power stations are now more expensive to build and more costly to run than wind turbines, solar farms, hydroelectric schemes, or fossil-fuel-burning power stations. However, along with the fossil-burners, nuclear power stations do provide a reassuring consistency not found in other sources of electrical production. Bear in mind, too, that the emotional reaction among the public to the concept of a waste-producing nuclear power station usually trumps the facts when it comes to political decisions.

Adjacent to the Ukrainian Chernobyl nuclear power plant, which failed in 1986, and caused the worst nuclear disaster in history, the abandoned city of Pripyat serves as a warning to nuclear engineers about the need for high levels of reliability and safety.

NUCLEAR PROPULSION

Since the dawn of the Atomic Age, nuclear power sources have been seen as a potential means of propulsion—on sea, in the air, and in space. Early development of nuclear power systems provided the US Navy with the world's first nuclear-powered submarine, the USS *Nautilus*, which in August 1958 became the first ship to sail under the North Polar ice cap and emerge after a submerged duration of 96 hours, having traversed 1,830 miles (2,940 km). Soon to follow were the first nuclear-powered submarines carrying submarine-launched ballistic missiles (SLBMs) and successive developments that today power all SLBM forces as a component of the nuclear triad involving land, sea, and air strategic nuclear forces. Nuclear reactors also power all 11 aircraft carriers deployed by the US Navy, and several ships from other countries utilize nuclear power for propulsion, including Russian ice-breakers and some naval vessels.

While nuclear power provides almost limitless electrical power for military submarines, and even those that carry nuclear ballistic missiles, it is also invaluable for long-duration research vessels such as the US Navy NR-1 submersible, here seen at Port Canaveral, Florida.

In all these applications, the overriding advantage is that nuclear-powered vessels are more autonomous in operation since they do not need frequent access to hydrocarbon fuels (useful when operating far from shore bases) and they have greatly extended mission life away from port. However, they pose unique difficulties for nuclear engineers. This is because marine reactors bring severe challenges both in the reduced area in which the equipment can be deployed and the harsh operating conditions, involving pitching and rolling in rough seas and the salt-water environment. Where a land-based power reactor would generate several thousand megawatts of electrical power, a marine power system is designed to deliver a few hundred megawatts at most. Due to its smaller size, the technical design of a marine reactor requires a much more highly enriched fuel, since the probability of a neutron escaping into the shielding before intersecting a fissionable nucleus is very much lower. It therefore contains a much higher percentage of U-235 than U-238, and this increases the probability of achieving fission. Proportionality varies, but most US vessels run on 96 per cent ^{235}U. Ships with lower-grade enrichment require refueling more frequently.

Marine nuclear power systems demand greatly increased safety and reliability features, which drive several innovative and design characteristics. For instance, while most land-based power plants use uranium dioxide fuel, marine reactors employ a metal-zirconium alloy, and with a long core life they incorporate materials with a high neutron absorption cross-section. The "poison" in the fuel elements is dissipated over time so that as they age, the fuel elements become less reactive. The life of the compact reactor is further enhanced by incorporating a neutron shield to reduce the damage on structural steel through bombardment.

For a brief period during the 1950s, some tests were conducted on a nuclear power plant for large military aircraft, believing that it would enable them to stay in the air for several days or weeks at a time, the aircrew accommodated in living quarters rotating their work shifts at the controls. At a time when tensions between the USA and the Soviet Union were at their height and the Russian space program achieved several "firsts," the US Air Force pushed the notion of a permanently airborne nuclear deterrent flying along the border with Russia. It was perverse logic that soon died, along with the plausibility of the concept. The engineering challenges of such an airborne reactor, and the hazard posed by an unfortunate accident, were deemed insurmountable and less attractive than more readily available means of flexing a deterrent.

Not so outrageous, however, were studies into a nuclear propulsion system for upper stages on launch vehicles. The US space agency NASA took a world

One attempt was made in the 1950s to test out a nuclear-powered aircraft, the equipment being installed in a modified B-36H photo-reconnaissance bomber and designated NB-36H. The reactor was flown but never operated.

lead in developing ***cryogenic propulsion systems***—the use of hydrogen and oxygen as propellants for producing high-energy combustion and a greater efficiency in the overall performance of their rockets. In a chemical rocket motor, the combination of these two propellants produces the highest efficiency that it is possible to obtain. A combustion engine heats hydrogen (the fuel) through its contact with oxygen at high temperature. If a nuclear reactor is used to heat the hydrogen to the same temperature instead, the oxygen need not be carried at all, saving weight and allowing more fuel to be carried. In this way, payloads of twice or three times the size of those lifted with a chemical combustion stage could be achieved. Or, with a similar payload mass, journey

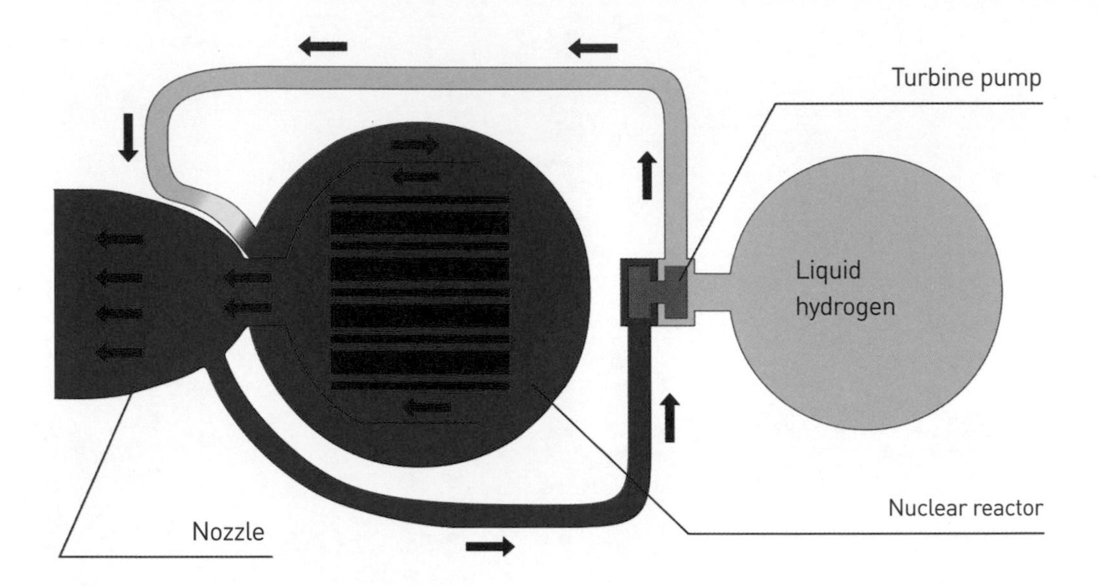

The operating principle of a nuclear thermal rocket, in which the oxidizer and combustion chamber is replaced by a reactor, raising the temperature of hydrogen fuel for acceleration and exhausted through a convergent/divergent nozzle for phenomenal efficiency.

times to distant planets could be reduced to one-third the trip time achieved with a conventional launch vehicle.

Nobody has ever suggested that nuclear rocket motors could be in the first stage of a launch vehicle; the contamination from a ground-launched thermonuclear rocket would be too great. But in the reaches of near-Earth space, where an upper stage would fire to accelerate a payload on its way to a destination in deep space, such a design is theoretically possible. However, the overall risks of launching a nuclear upper stage from the ground and the public reaction to such a vehicle have been assessed as too great to justify development. Instead, reduced trip times and faster velocities could be achieved with solar-electric motors, as we will see in Chapter 13.

Harnessing the atom and using it for producing electricity has been the hallmark of nuclear engineering since the end of World War II in 1945. That same knowledge has been used to build engines that may power the spaceships of the future. But electrical production and electronic engineering have made equally impressive progress, powering the modern world.

Chapter Ten

ELECTRICAL AND ELECTRONIC ENGINEERING

Early Evidence of Experiments with Electricity—The Unifying Forces—The Characteristics of Electricity—Delivering Electricity—The Rise of Electronic Engineering—The Relevance of Electronic Engineering

PEOPLE

CURRENT
|
AMPS
|
VOLTAGE

GEORG OHM

THALUS OF MILETUS

MICHAEL FARADAY

FRANCIS HAUKSBEE

JAMES CLERK MAXWELL

ELECTRICAL

ELECTRONIC

CHRISTIAN HÜLSMEYER

ELECTROMAGNETISM

APPLICATIONS

RADIO

RADAR

EDISON MACHINE WORKS

TESLA AC SYSTEM

ELECTRICITY POWER GRID

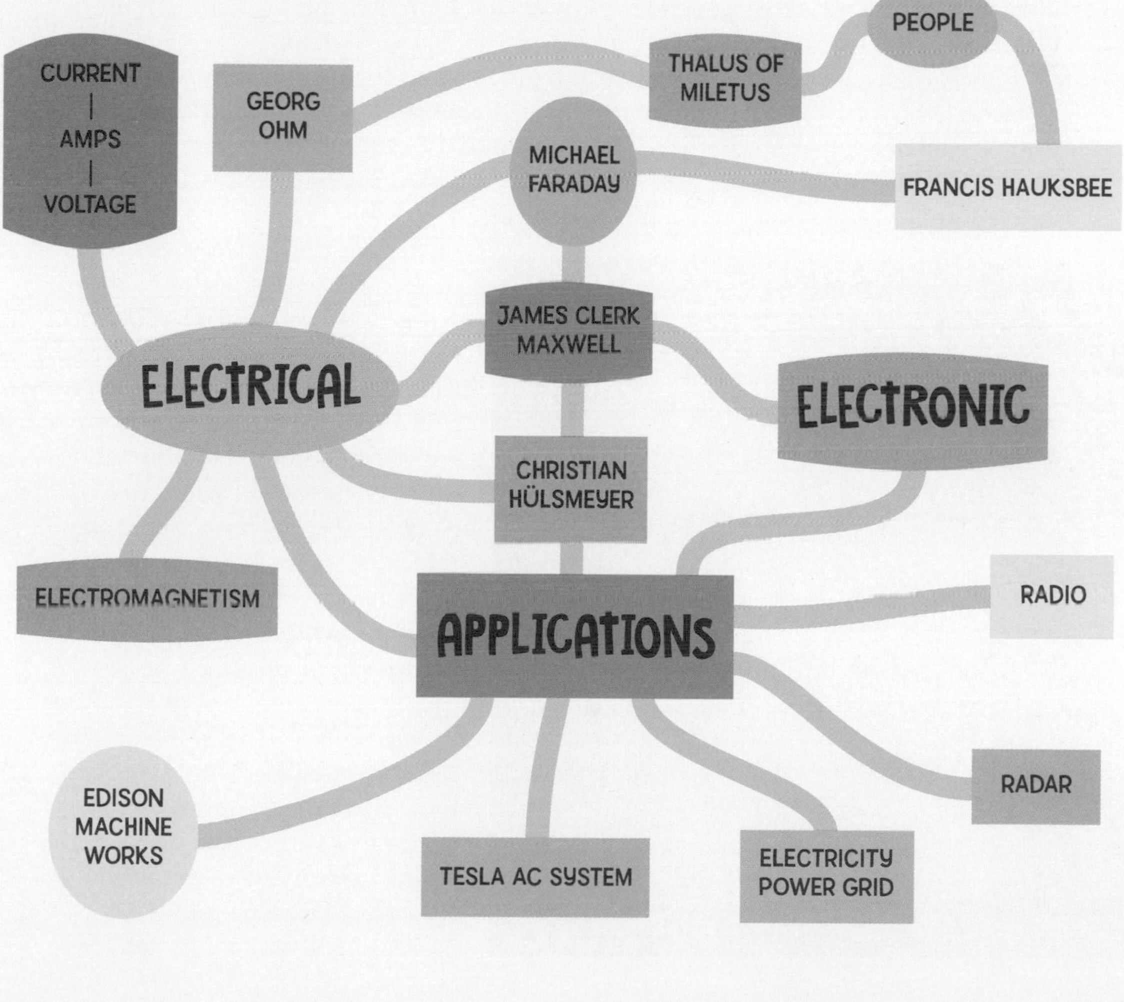

EARLY EVIDENCE OF EXPERIMENTS WITH ELECTRICITY

So far in this book we have been firmly attached to the origin, development, and application of mechanics and the principles upon which simple and complex machines have been built and implemented in a wide range of applications. From the Stone Age to the Space Age, machines have helped to create the modern world through a progressive series of steps, each building on the one before. Some inventions and innovative technologies have been game-changers, such as the steam engine, the internal combustion engine, and the nuclear industry, in terms of both power production and atomic weapons. And then there's one of the most all-embracing developments: the understanding of electrical energy and how it could be engineered into useful machines. In this chapter, we review the origin and evolution of the underlying principles behind electrical and electronic engineering.

Once again, we look to the classical world to find the origins of how people became aware of electricity and conducted experiments with electric shocks, and started to explore the nature of lightning and the way electrical energy could be applied to practical uses. The 7th-century BC Greek philosopher Thales of Miletus (*c.*624–*c.*548 BC) understood static electricity and tested the properties of attraction between fur and other materials when they were rubbed together, relating that phenomenon to the sparks that could be observed when two pieces of amber were rubbed together. This built on the well-recorded accounts of electric shocks that could be received from various types of fish and rays in the sea, data for which go back to the beginning of recorded times.

One of the mysteries of the ancient world, lightning has fascinated humans for millennia, speculatively inducing belief in mystical powers, gods, and the administering of life and death, killing humans and livestock and starting fires.

As we have seen throughout the history of mechanical engineering, clear and deductive thinking powered the empirical age of learning through natural logic and simple experiments—processes of observation rather than preconceived assumptions. So it was with the ancient world when it came to electrical forces, as explained by the Greek philosopher Democritus (*c.*460–370 BC), who said that all matter was assembled from what he called "atomos," microscopic units of matter so small they were impossible to see which formed the basis of all physical things and "inexplicable forces." There is some evidence from 1500 BC, albeit disputed, that the electrical properties of certain materials were used to create a simple battery and that the Hebrew Ark of the Covenant was such a device. Some of the properties and descriptions from ancient Hebrew and Mesopotamian texts closely resemble the so-called Baghdad Battery, an assemblage of items known to be more than 1,500 years old that would, if employed properly, allow ***electroplating*** to be carried out. Such an interpretation was proposed as an explanation for ancient objects found in modern-day Iraq that have a gold coating on silver objects, inferring a use of this method.

THE UNIFYING FORCES

The scientific study of electricity really begins in the mid-18th century when the English scientist Francis Hauksbee (1660–1730) inadvertently discovered the principle of what would later become the ***gas discharge lamp***—a means of generating light by sending an electric discharge through an ionized gas (or plasma). By placing a small quantity of mercury in a glass from which air had been removed he was able to obtain a modest glow as a charge caused by rubbing his hand on the globe. He did not understand the physics but he did observe and record the results. So too did Benjamin Franklin (1706–90), the political philosopher and one of the founding fathers of the American Republic, who became obsessed with the peculiarities of electricity, to the extent that he sold personal possessions to fund his research. In particular, Franklin became fixated on the properties of the ***Leyden jar***, a device for storing electrical energy. It consisted of a glass bottle half-filled with water, with foil attached to the inside and outside surfaces and a metal terminal projecting through the lid stopper to make contact with the water. As a demonstration of conduction, the Leyden jar was the first capacitor, and is still used today to educate students in the principles of electrostatics.

Significant progress leading to the current understanding of electricity took hold during the 19th century. It began in 1827 with research by the German scientist Georg Ohm (1789–1854), who would give his name to the difference between the potential (voltage) across the conductor and the resulting electric current—***Ohm's law***. In 1831, Michael Faraday (1791–1867) discovered what today we know as ***electromagnetic induction***. By demonstrating electromotive force across a conductor in a varying magnetic field, Faraday was jumping ahead of his day.

Well known as one of the first to make direct measurements of electrical energy by the use of a Leyden jar, Benjamin Franklin is remembered for his daring experiments with lightning.

The 19th century would see the definition of different forces, with electricity and magnetism regarded as separate until in 1873 James Clerk Maxwell (1831–79) convinced scientists to think of the combined force of electromagnetism as fields defined by his mathematical equations. Just five years later, James Wimshurst (1832–1903) began working on a machine for developing high voltages and this would carry his name. Consisting of two Leyden jars and two contra-rotating glass discs with metal strips, it comprised two metal bars and metallic brushes with a spark gap between two metal balls.

It took very little time for the development of electromagnetic theory and experimentation to spawn a new field of academic learning, and in 1882 in Germany, the Darmstadt University of Technology founded the world's first faculty for electrical engineering. Around the same time, in the USA, the Massachusetts Institute of Technology (MIT) started running courses in electrical engineering in the physics department. In this century of invention and engineering applications, the use of electricity for commercial purposes grew

The discovery of the Leyden jar occurred as a result of work in the laboratory of Pieter van Musschenbroek (1692–1761). In 1745, Andreas Cunaeus (1743–97) duplicated the experiment by rotating a glass sphere, believing he was creating static electricity. A charge was conducted to a glass filled with water into which was suspended a bar on a chain. Cunaeus felt a powerful shock when he touched the chain. Acting as a capacitor, the water built up a large charge greater than that possible from an electrostatic charge, an equal but opposite charge building up in his hand. He took several days to recover.

rapidly. ***Street lighting*** began to appear during the 1870s and its application widened with the advent of ***arc lamp technology***, providing electric lighting for the public on an expanding scale. Such lamps produced an early, if primitive, means of creating light through an arc between carbon electrodes in air inside a glass bowl. By the 1880s, the ***transformer*** had arrived, a means of transferring electrical energy from one circuit to another. This occurred at about the time that Thomas Edison (1847–1931) began supplying the first public utility in 1882 with a 110-volt ***direct current*** supply in which a single, or unidirectional, flow can be made to travel along a wire—a ***conductor***. But battlegrounds lay ahead.

Edison held firm to his belief in the direct current (DC) system, but a potentially commercially competitive system arrived with the ***alternating current*** (AC) system, proselytized by Nikola Tesla (1856–1943) of the then Austro-Hungarian Empire (now a region in Croatia). Having moved to the USA in 1884, Tesla worked for a while with Edison Machine Works before leaving to set up on his own, whereupon, with the help of friends and financial assistance, he developed the AC system. This was further advanced with the ***polyphase*** patents, whereby three or more energized conductors carry alternating currents with a defined phase angle. These are especially useful for electric motors that use AC power to rotate.

A pioneer of electrical engineering, Michael Faraday, from a painting by H. W. Pickersgill.

Michael Faraday in his laboratory at the Royal Institution, painted by Harriet Moore.

The reconstruction of Thomas Edison's Menlo Park laboratory at the Henry Ford Museum, Dearborn, Michigan, USA.

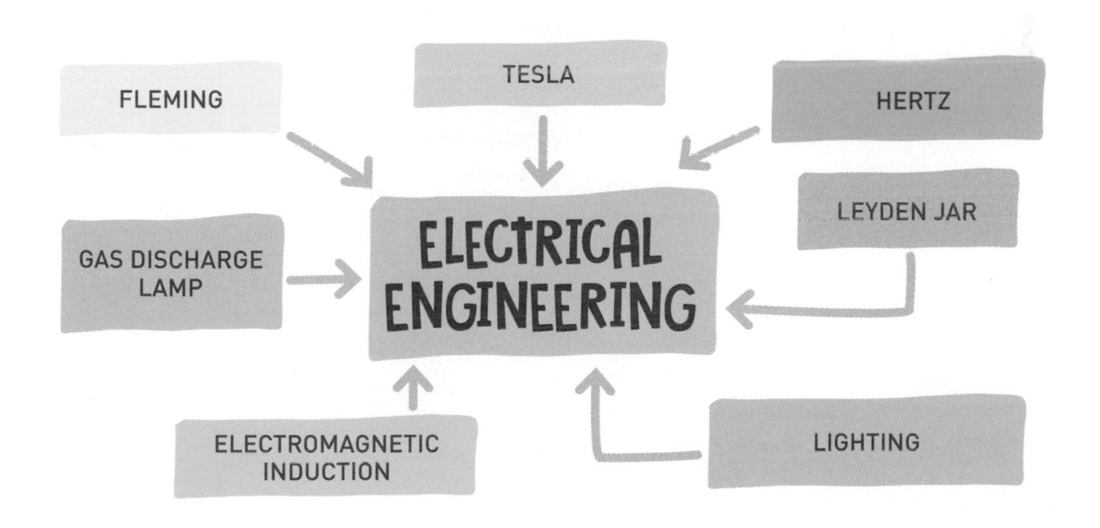

The difference between AC and DC systems
In an AC system, the current changes direction at a frequency measured in Hertz, the US supply providing about 120 volts at a frequency of 60 Hz, which means that the current switches direction 60 times per second. The DC system, which doesn't switch direction, is frequently used for batteries and solar cell supplies.

However, it was Westinghouse Electric Corporation and not Tesla that developed the AC system, and by 1888 the company had 68 AC stations to Edison's 121 DC plants. A complex and highly contested battle erupted between advocates of each system, with an increasing numbers of deaths from the AC system bringing the concept into disrepute and creating alarm among the public. Accusations of commercial plagiarism and theft of intellectual property inflamed the debate into a rancorous argument that was taken to the courts, and the escalating number of deaths from high-voltage AC systems raised for a while the entire question of its implementation. The problem was that although the AC system was cheaper and more efficient than the DC system, it was some time before it was adopted universally and a safe supply was developed. In the interim, bizarre tests were conducted to establish the appropriate level of power that should be used for the legalized electrocution of convicts who had received the death penalty. Eventually, a level of 1,000–1,500 volts was settled on following tests involving the electrocution of horses.

THE CHARACTERISTICS OF ELECTRICITY

Electricity is essentially a ***flow of charged particles*** and because in metals the electrons are free to move, they are regarded as good ***conductors***. Materials that do not conduct electricity are useful as ***insulators***. Through these principles, an electric charge can be either ***positive*** or ***negative***. The movement of the charge is an electric current and produces a ***magnetic field***. The magnitude of the electric force is found by Coulomb's law, which defines the amount of force between two electrically charged particles. The amount of force between two charged bodies at rest is referred to as the ***electrostatic force***.

The importance of Coulomb's law
This law, which was published by the French physicist Charles-Augustin de Coulomb (1736–1806) in 1785, was the origin of the theory of electromagnetism and as such was described in one of the most important papers that contributed to the overall work developed by Maxwell.

Defining units of measurement

Definitions and terminology and the characteristics of a power unit are defined by units and symbols, of which the most basic are ***voltage*** (volts), ***current*** (amps), ***resistance*** (ohms), and ***power*** (watts). Wattage is always equal to voltage multiplied by the current and a simple analogy can help guide definition, where the various units can be compared to a water system: voltage is the pressure of the water supply, the current is the rate of flow, and the resistance is the size of the delivery pipe. Electrical engineering unifies voltage (V), current (I), from the French *intensité du courant,* and resistance (R) through Ohm's law, which states that the voltage is equal to the current flowing in the circuit multiplied by the resistance: $V - I x R$. Thus will it be seen that an increase in voltage will also increase the current and if the resistance is increased, the current will decrease.

Volts

Named after the Italian inventor Alessandro Volta (1745–1827), the formal definition of a volt is the "difference in electrical potential between two points of a conducting core when an electric current of one ampere dissipates one watt of power between those points." The source of the electrical energy provides the potential difference in the circuit, the voltage, and this is referred to as the ***electrical pressure voltage***—measured in volts. The current increases as the number of volts increases, but the flow of the current requires the conductor, or wire, to loop back to the source. If that circuit is broken, by a circuit breaker or a switch, then the flow will stop. In complex machines, this can be the method by which circuits can be opened or closed, usually by automated, programmable, or autonomous control.

Amps

The amp, or ampere—as named after French physicist André-Marie Ampère (1775–1836)—is the root unit in electrical engineering and is the internationally recognized SI unit. It is defined as "that constant current which, if maintained in two straight parallel conductors of infinite length, or negligible circular cross-section, and placed one metre apart in a vacuum would produce between these conductors a force equal to $2x10^{-7}$ newtons per metre of length."

Ohms

The ohm is an SI unit which is a measure of electrical resistance named after Georg Simon Ohm. It is the electrical resistance between two points of a conductor when a constant potential difference of one volt produces a current of one ampere in the conductor.

Watts

Named after the Scottish engineer James Watt (1736–1819), 1 watt (1 W) is the base unit of power in electrical systems but it can also be used in mechanical systems. It measures the quantity of energy that can be liberated per second by any particular system. The watt is defined as the energy consumption rate of one joule per second: 1 W = 1 J / 1s. It can also be defined as the current flow of one ampere at one volt (1 V): 1 W = 1 x 1A. In a variation, ***peak-watts*** (Wp) is the term used to define the maximum amount of energy that can be obtained from a ***photovoltaic*** (pv, or solar) cell and is usually the product of an industry-approved test since that is frequently used for defining the specification by which pv cells are marketed.

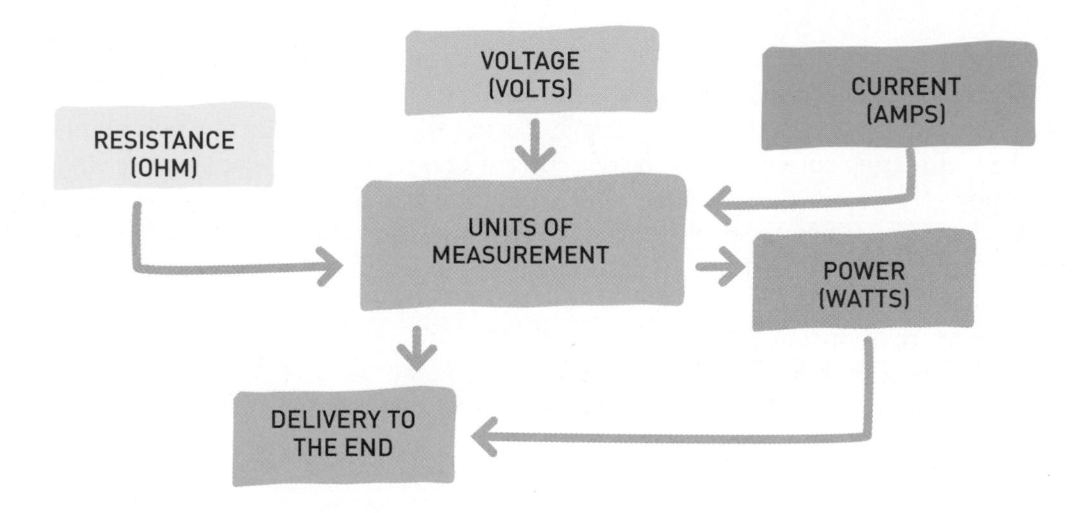

DELIVERING ELECTRICITY

There are various ways in which different questions can be answered using a standard set of equations. For instance, to convert watts to volts $(_V)$, the voltage is equal to the power $(_P)$ in watts $(_W)$ divided by the current $(_I)$ in amps $(_A)$: $V(_V) = P(_W) / I(_A)$. To convert watts to amps, the current is equal to the power in watts divided by the voltage: $I(_A) = P(_W) / V(_V)$. Furthermore, to convert watts to ohms: $R(\Omega) = V(_V)^2 / P(_W)$. For engineers interested in the amount of energy available over a given duration, one ***kilowatt-hour*** (expressed as kWh) is the energy consumed by the power consumption of 1 kW in one hour.

All these factors come into play when determining the most effective system to use for a mechanical device, or the most efficient one when designing and constructing a national power grid, for instance. And here it is helpful to go back and look at the difference between AC and DC systems. An AC system can be displayed on an oscilloscope and show a sine wave oscillating between peaks of -170 volts and +170 volts around a mean but with an effective power output of 120 volts—the standard grid voltage in the USA. The observed oscillation rate is 60 cycles per second (cps) and as we have seen, this is why it is called an alternating current. Power generators generate AC power naturally, and so conversion to DC requires an additional step, but since transformers are essential in the power grid anyway and it is easy to convert AC to DC, this isn't necessarily a problem. However, it is expensive in terms of energy to convert DC to AC, and this is essentially the reason why Edison found it commercially productive to have a power system based on AC.

But that's the big picture, and the choice between the two systems provides different options for smaller machines and domestic devices. Characteristically, an AC device will have a voltage rating in excess of 100 volts whereas a DC device will have a lower rating, usually of less than 100 volts. In terminology, the current and the voltage can be written together, so that a device rated at 120 V could be written as 120 VAC. As an example, an AC electric fan will have a voltage rating of 11, 115, 120, 220, 230, or 240 volts, whereas a DC fan will have a rating of 3.5, 12, 24, or 48 volts. Powered directly through a utility outlet or a transformer (step-up or step-down), the AC device would have a die-cast aluminum case whereas a DC fan, powered through a transformer or from a battery, would have a case made from thermoplastic.

Another factor comes into the bigger picture: the three-phase power supply that comes out of a power station. Three separate phases provide an optimum consistency of power output, effectively a "smoothing" of the sine wave. In one- and two-phase power there are 120 moments each second where the wave crosses 0 volts, but in three-phase power, which is optimized for both industrial and domestic use, the peak evenness of power output is achieved. In four-phase power supply, the advantage diminishes dramatically and it is therefore not economical because the difference is so small and has no effective return for the added cost of laying out a four-phase system.

Because of this, a transmission substation is located outside the power station's generator in order to step up the output to extremely high voltage for transmission along the grid, and this can be approaching 800,000 volts. Of course, resistance scavenges less energy throughout the length of the transmission line, which can be up to 310 miles (500 km) along the towers. The three tower wires carrying the three phases are usually accompanied by additional lines to attract lightning and disperse it to ground. For general use at engineering facilities—and even the domestic home distribution network—the power is stepped down from the transmission side to the distribution outlets, and that can occur at a power substation where the voltage is reduced, typically from 500,000 volts to 50,000 volts and thence to a distribution voltage of less than 10,000 volts. From there, it is common for two separate distribution lines at two different voltages to spin off into

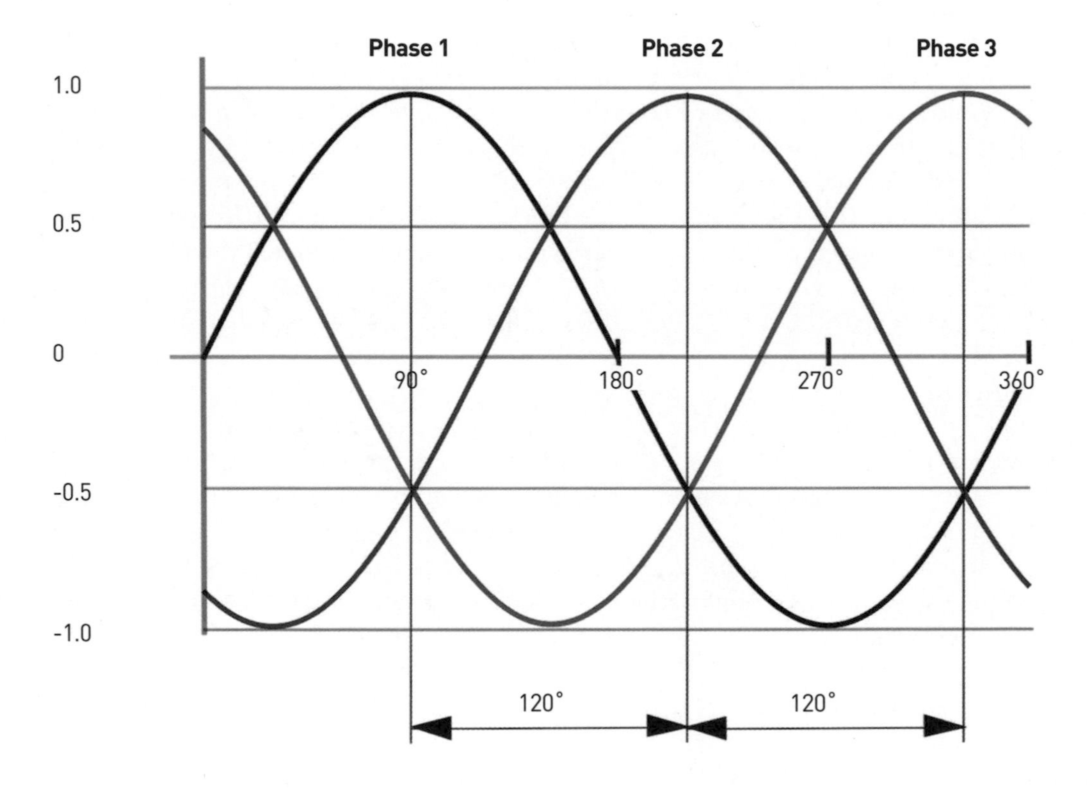

A three-phase alternating current waveform. This figure illustrates one voltage cycle of a three-phase system, labeled 0° to 360° (2 π radians) along the time axis. The plotted line represents the variation of instantaneous voltage (or current) with respect to time. This cycle will repeat 50 or 60 times per second, depending on the power system frequency. The colors of the lines represent the American color code for a 120-volt three-phase system.

multiple distribution points, but for domestic use these will have to be stepped down again. This action is usually assigned to dedicated substations or smaller transformers. For some domestic use, power is regulated to prevent surges or under-voltage conditions, and because homes and small businesses can quite easily operate with only one- or two-phase power, taps are run off for down-phasing the supply.

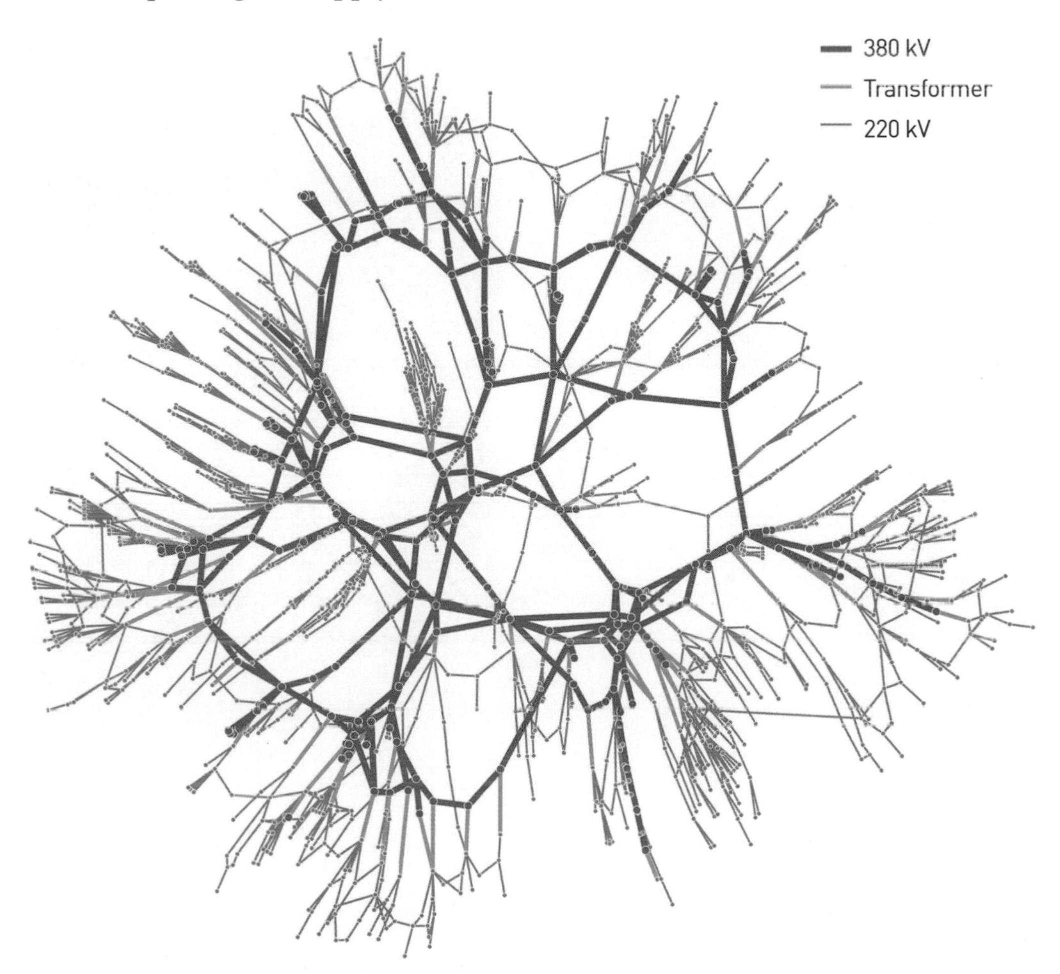

A high-voltage transmission system illustrating the way in which different power levels are connected. The diagram shows how spurs and trunks are routed. It does not imply a geographic layout.

The monument in Edinburgh, Scotland, to James Clerk Maxwell, the father of electromagnetic theory and of electrical engineering.

Sample cross-section of a typical conductor for a 500 kV DC overhead power line on the Inter-Island transmission system in New Zealand.

THE RISE OF ELECTRONIC ENGINEERING

While mechanical and electrical engineering are proudly the product of the Age of Enlightenment and the scientific revolution, which brought unimaginable strides in human progress, electronic engineering emerged through commercial applications of the research into the laws of electromagnetism in the latter half of the 19th century. For example, the development of radio grew out of electrical telegraphy, employed from the 1840s as a point-to-point messaging system, which spawned a wide range of telegraphic mechanisms. These helped greatly with the development of British interests around the world and in opening up the Great Plains and the West of America as the USA expanded towards the Pacific coast.

Perhaps it may be seen as astonishing, yet only to be expected, that in the decade when Americans were slaughtering each other during the Civil War, the greatest discovery of modern times—and probably for the next several hundred years—was taking place: the unified theory of electromagnetism. In importance, it was every bit the equal of Newton's unifying principles of gravity. Between 1861 and 1865, James Clerk Maxwell's theories were developed and published, giving inspiration to scientists and inventors around the world. While the properties of electricity were made practical for exploitation through the unified effects of magnetism, wireless communication by radio signal was arguably the second-most important development. The discovery that the electromagnetic spectrum can be expressed in frequency, wavelength, or photon energy opened up infinite possibilities ripe for practical, and commercial, exploitation.

The four forces

The electromagnetic spectrum embraces a wide range of ***frequencies*** (the number of occurrences of a repeating cycle in a given period of time), the ***wavelength*** (the distance over which the shape of the wavelength repeats itself), and the ***photon energy*** (that carried by a single photon). Over time, ***electromagnetism*** (the interaction of particles with magnetism) would be seen as the fourth of the four known forces; the others being ***gravity*** (the weakest of all but the only one capable of multiplying at a rate proportional to the mass), the ***weak nuclear*** (radioactive decay), and the ***strong nuclear*** (the binding force that keeps protons and neutrons together in a nucleus). However, more recently, in the 1970s and 1980s, electromagnetic forces and the weak nuclear force have been found to consolidate into a single electroweak interaction mediated by a ***boson***, a class of subatomic particles that are believed to carry the four known forces.

Indeed, the next step forward in the technological application of electromagnetism was stimulated by purely commercial incentives. Along with theoretical work by Oliver Heaviside (1850–1925), the four vector equations written by Maxwell set down the principles upon which *radio*—point-to-point communication without the aid of a wire—was made possible. History generally regards Guglielmo Marconi (1874–1937) as the "inventor" of radio—and indeed in 1896 he demonstrated the first practical application of communication by radio waves—but that would be to put too much credence

Diodes and triodes

Limited means of wireless communication had been demonstrated in the last two decades of the 19th century, and even Tesla got in on the act. The first ***vacuum tube*** was invented by the British engineer John Ambrose Fleming (1849–1945) in 1904, and was effectively a ***diode*** in which a two-terminal electronic component delivers high current in one direction with low resistance at one end and high resistance at the other. Two years later came the ***triode***, with three electrodes inside a vacuum jar consisting of a heated filament (a ***cathode***) where a conventional current leaves the device, a grid, and a plate (an ***anode***, or positive electrode) where the polarized current enters the device. The triode really began the field of electronics.

on the application of principles laid down by others; Marconi stitched the connections, designed and developed the equipment, and demonstrated a workable radio communication system, but there have been several claimants and contested attribution only fuels the uncertainty. Nevertheless, despite the haziness of its origins, from radio came the application of radio-direction-finding (RDF), or *radar* for short.

Radio and radar played a highly significant part in electrical and mechanical engineering during the 20th century, and the development of this technology since the 1930s has overtaken many other industries, increasingly and significantly challenging the "wired" world in both civil and military applications. The origins of radar can be traced to before World War I, being found in the theoretical writings of several eminent, and not so well-known, inventors. The principles were established by Rudolf Hertz (1857–94) in the second half of the 19th century through the application of practical experimentation, demonstrations, and tests that had not been possible in the days of Maxwell. Hertz demonstrated through a spark transmitter a means of sending radio signals at a frequency of 50 MHz and followed that by showing that they could be reflected, refracted, diffracted, and polarized, just as was shown to be possible with light waves. Hertz, however, could find no practical use for all of his discoveries, and it was left to others to apply the physics and bridge the gap to electrical engineering.

Thus, although Hertz did test the use of metal sheets to reflect radio waves and so become the first to demonstrate the principles of radar, it was the German Christian Hülsmeyer (1881–1957) who lodged the patent for the first practical radar device on April 30, 1904. This was partly due to th e fact that he had been traumatized by hearing a small child's dying screams as it was crushed between two barges that had scraped together in fog on the Rhine river, and in response he had turned to the theories of Maxwell to find a way to design a device that could give warning of approaching barges in the worst weather conditions. Hülsmeyer duly came up with a device with a simple spark gap to generate a signal aimed at a target with a cylindrical parabolic reflector, a separate coherent receiver picking up the reflected echoes. Called the *telemobilscope*, the device was sent to potential clients but few, including the German Navy, could see any use for it! Later, modifications would allow it to display the range to a target. As we saw in Chapter 9, radar was subsequently developed into a fully operational air defense system by the British in the late 1930s and while, by that time, Germany was at the forefront of radar development, it was the RAF that had the benefit of the world's first fully operational early-warning technology.

THE RELEVANCE OF ELECTRONIC ENGINEERING

It was largely a result of the development of the telegraph, radio, and telephone industries that the field of electrical engineering arose and produced new applications that underpinned new possibilities. Equipment to expand these opportunities gave birth to the pure electronics industry, and along with that came the ***transistor***, the ***semiconductor,*** and s***olid-state electronics***. Collectively, these enabled a wide and ever-expanding range of applications, new technology, and a level of interdisciplinary specialities that did a lot to unite research teams working across several applications in a variety of experimental and commercial activities.

For example, without electrical engineering and the fields of radio and radar, ***telemetry***—the means to collect and transmit data acquired from a range of sensors on moving machines at sea, in the air, and across space—would not have been possible, and it would also not have been possible to send information vital to monitoring performance, reliability, and for controlling operations. In fact, the space program would not have been possible without the ability to record, transmit, and interact with signals sent from Earth and, from the receiving device in space, back to ground stations. Looking to the future, and closer to home, electrical energy, which we will explore in the next chapter, is a possible solution to the problem of polluting aircraft engines that operate on hydrocarbon fuels. Beyond these more immediate challenges of developing environmentally friendly and carbon-neutral technologies, the very possibility of artificial intelligence (AI) is also only conceivable thanks to the microelectronics and highly developed computer systems that have evolved out of the age of electronics.

Chapter Eleven

POWER PRODUCTION AND GENERATION

It's an Electric World—Dynamos and Alternators—Storage Systems—Internal Combustion Engine Vehicles—Electric Vehicles—Fuel Cells—Fuel Cells of the Future

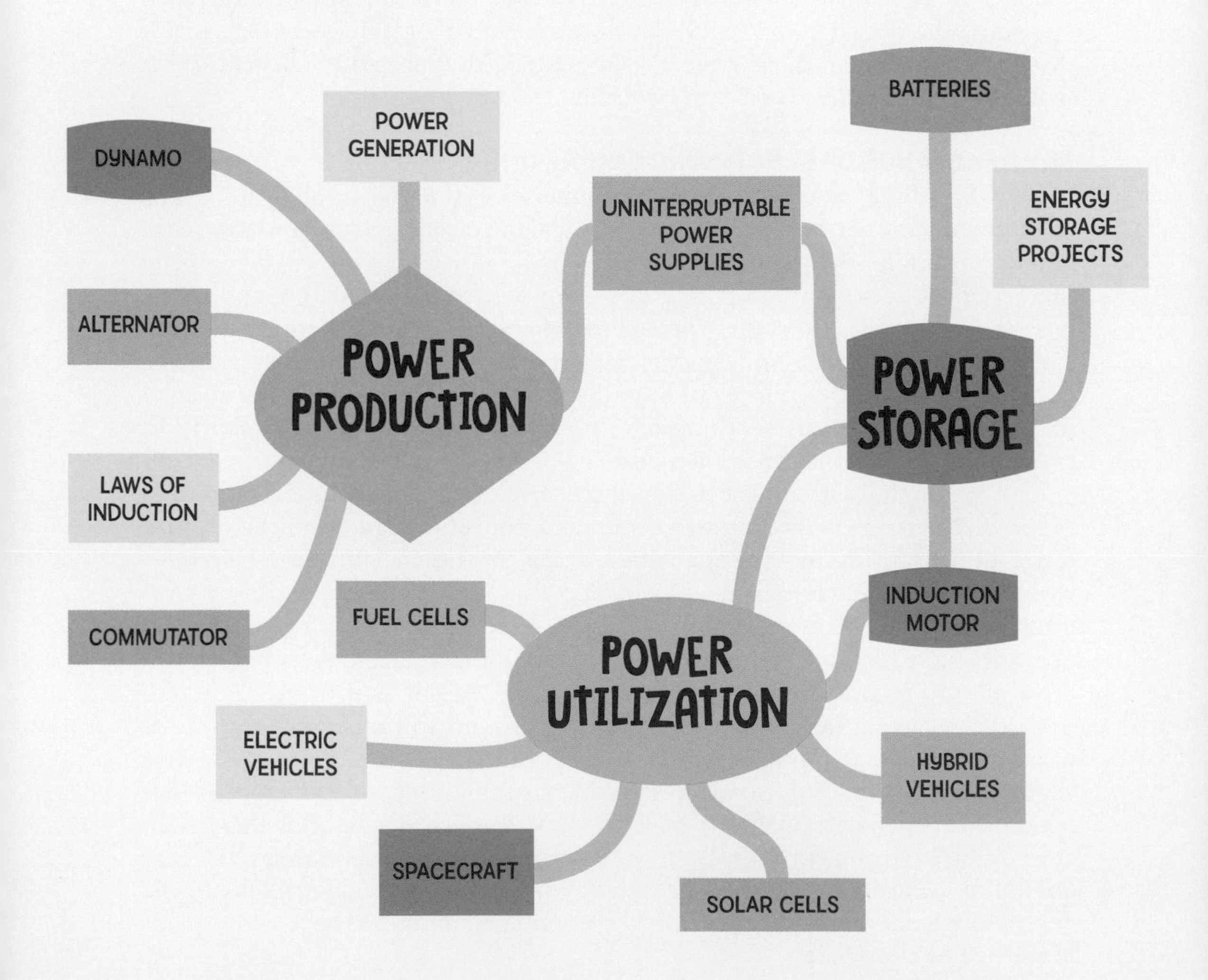

IT'S AN ELECTRIC WORLD

There is nothing more fundamental to the modern world than the production and distribution of electricity. It powers everything from cell phones to cars, and from domestic lighting to household white goods. A consistent supply of electrical energy is vital for a wide range of medical equipment and for functions as diverse as air traffic control systems and emergency services, and more trivial pursuits such as toys, playthings, and entertainment. We cannot, it seems, do without electricity. For the modern economy, electricity powers society at a domestic, business, and industrial level, with greater demands on electrical power supplies driven by the shift from hydrocarbon fuels to electrical energy for both static and mobile machines, including cars, trucks, and trains.

Having examined the fundamental principles of mechanical engineering in earlier chapters, and introduced the subject of electrical and electronic applications, we must now examine the use of mechanical principles to the need for generating electrical energy. We will limit ourselves to the mechanical

A turbo-generator produced by Siemens in Germany.

processes of producing power distributed on national and local grids and networks, leaving mechanical processes involved in producing electricity from solar photovoltaic cells and by geothermal means to the next chapter. Finally, because engineers are at the forefront of discovering what is possible, we will look at the way we can plan for a more sustainable future using electrical energy sourced from technology that can reduce the human impact on the environment.

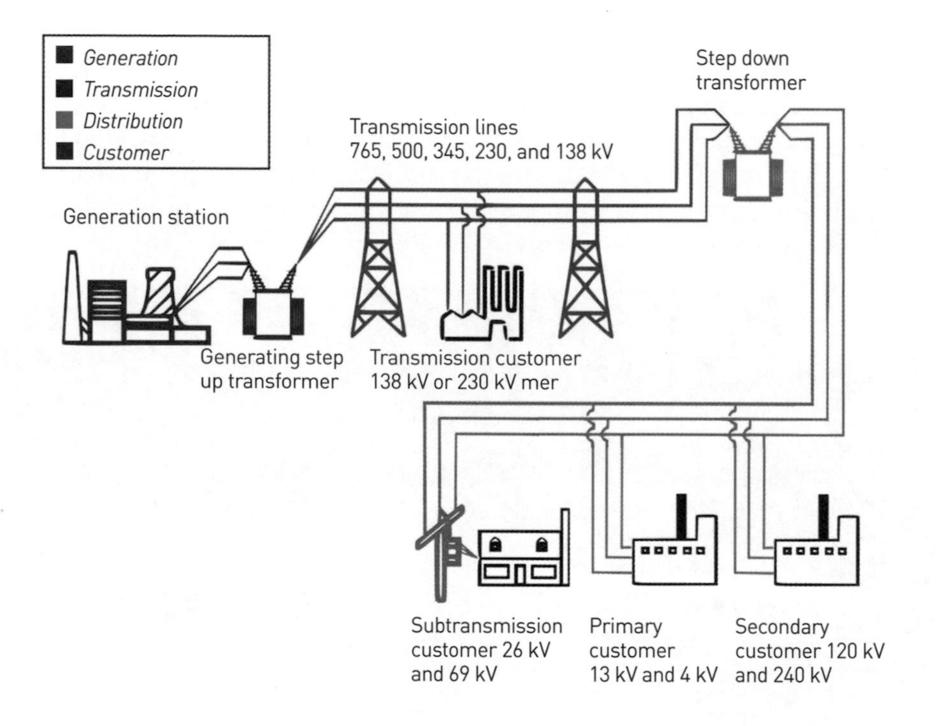

DYNAMOS AND ALTERNATORS

There are two ways of generating electromagnetic energy: the ***dynamo*** and the ***alternator***.

The dynamo

This device uses coils of wire and a magnetic field to convert rotary motion into a direct current, as laid down by ***Faraday's law of induction***, which predicts how a magnetic field will interact with an electric circuit to produce an electromotive force, or electromagnetic induction. The dynamo comprises a ***stator***, or static structure, which contains a constant magnetic field, with a set of rotating windings known as the ***armature***, which are made to rotate within that field. As explained by Faraday, the movement of the wire creates an electromotive force, pushing on the electrons in the metal and causing an electric current to run in the wire. Small machines can use magnets to provide the magnetic field but for larger machines requiring greater power, electromagnets are provided, sometimes known as ***field coils***.

From this alternating current, DC is enabled using a ***commutator***, which is really a rotary switch accommodating a set of contacts on a shaft with graphite-block stationary contacts known as ***brushes***. As it rotates, the commutator reverses the connection of the windings to the external circuit and produces a pulsed DC current. As explained in the previous chapter, AC power was believed to have little use, and the development of the dynamo was a means of replacing liquid batteries, which are difficult to accommodate in small machines. Dynamos, by contrast, are extremely useful for producing electric light via the rotary motion of the pedals for small lamps fitted to bicycles—an application for which the storage battery is of limited use thanks to its short life and lack of convenience—both of which are always a spur to innovation.

Early applications found use for dynamos in power stations generating electricity and were significant in replacing steam for this purpose. However, there are certain limitations with dynamos for producing electricity, not least the difficulty in disengaging windings from those in series to those in parallel, which have not had their mechanical drive systems linked in special ways at the rotor or field wiring levels. Nevertheless, there are still some uses for dynamos in low-power applications where it is preferable to have low levels of voltage, although they are fast disappearing in many of the traditional applications where they were once common, not least in motor vehicles, where they have been replaced with alternators.

The alternator

Alternators convert mechanical energy to electrical energy to produce alternating current. In this case, the armature is stationary while the magnetic field rotates. In other cases, as with a linear alternator, a rotating armature is placed within a stationary magnetic field. It is no longer common to use the word "alternator" to imply an AC electrical generator; that word is now almost exclusively retained for the small rotating machines used to create electrical power for cars, motor cycles, trucks, and buses—supporting relatively small internal combustion engines. Thus, the terminology changes over time, and an alternator that incorporates a permanent magnet for a magnetic field is now known as a ***magneto***.

As noted in the previous chapter, the majority of the world's electrical power is obtained from 50–60 Hz, three-phase alternating currents. During the early 20th century, one of the limitations on large power generators was the strength of the magnetic field provided by permanent magnets—a problem that was partially solved by what was known as a process of ***self-excitation***. For this, a small amount of electrical power from the generator was tapped off to an electromagnetic field coil, and this enabled the generator to produce significantly more power. To accomplish this, a small amount of ***remanent magnetism*** within the iron core provides the magnetic field to start the

This alternator was built by Ganz Works in 1909, installed in the generating hall of a Russian hydroelectric plant.

generator, producing a small current in the armature. As this flows through the field coils, connected either in series or in parallel with the winding, it produces a larger magnetic field, which in turn produces a larger armature current. This process continues until the magnetic field in the core levels off due to saturation and the generator itself achieves optimum, steady-state output. But even this is insufficient in very large power stations and it is not uncommon for a smaller, but separate, generator to excite the field coils in the larger generator. This can result in a dangerous situation during a universal power outage in that there is no power to energize the subsidiary generator to get the supply reinstated.

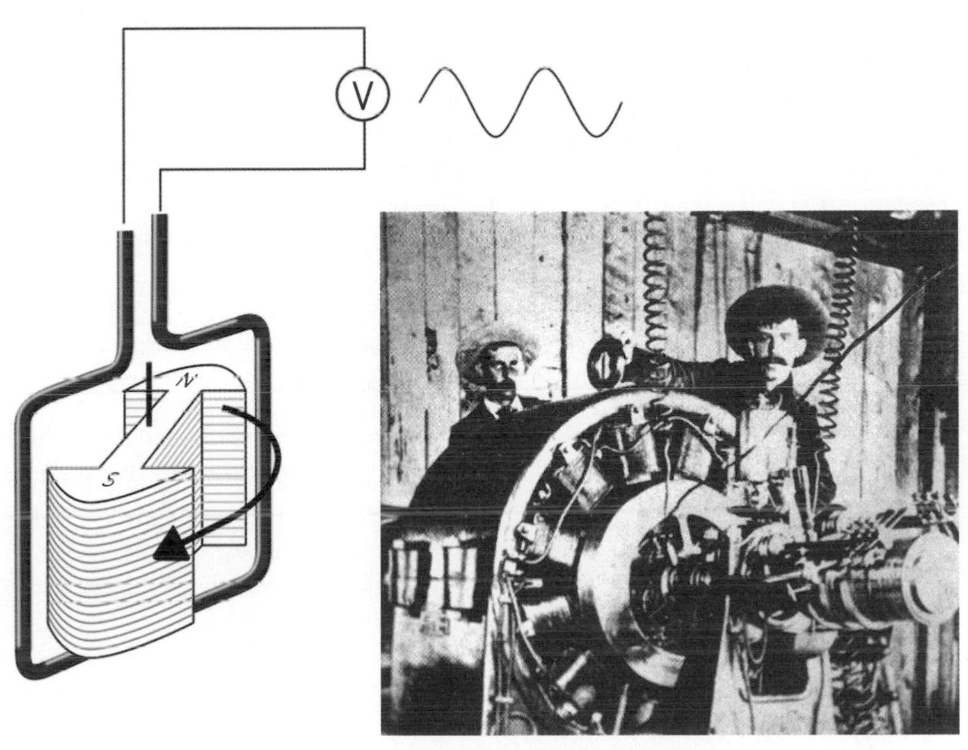

A highly simplified diagram of the operating principle of an alternator, showing a rotating magnetic core and the stationary stator, the current (v) being induced by the rotating magnetic field.

Credited with being the first industrial application of AC power, this Westinghouse synchronous alternator was installed at the Ames power plant, Colorado, in 1891. It was connected by belt to a water wheel, producing 3,000-volt, 133 Hz, single-phase power driving a similar alternator 4.2 km (2.6 miles) away, acting as a motor at a stamp mill at a mine. It dispensed with a steam mill, which was difficult to service due to diminishing quantities of timber.

STORAGE SYSTEMS

It is one thing to produce electricity and apply it to practical use, either in a domestic supply grid or a motor vehicle, but quite another to accumulate that energy and store it for future use. In its most basic form, a battery is an ***accumulator***. In advanced applications, batteries can themselves provide resilience against power failure at source or at intermittent stages of a grid system. At the lowest scale, devices categorized as providing an ***uninterruptable power supply*** (UPS) are available for small offices and work stations where a loss of power would be unacceptable. For example, UPS equipment is essential in hospitals where critical equipment must be kept running, and in larger applications, government servers, and integrated electronic systems for controlling product distribution. Whatever the setting, they provide assurance of reliable power connections.

Increasingly, where power supplies are not predictable, a more robust system is necessary, especially for alternative energy sources using unpredictable wind or wave power, for example. By controlling the load on a power distribution system, power storage systems can help with what is known as "peak shaving," controlling the load rather than the power output, with surges in demand being met by a supplementary load input. This is all part of consumer control made possible by peak tariff charges as an inducement to managing demand at the user end and smoothing out the curves on power take-up.

Type of battery

Various types of battery can be used for energy storage. The principle is not fundamentally different to that provided by the battery in a motor vehicle, except that certain types of battery are more suited to energy storage for the main power supplies to domestic and industry users.

Sodium-sulphur batteries are known to age with time and will be subject to random failure unless used frequently, while other battery types can suffer from repeated charge and discharge cycles, increasing at high charge rates. There are dangers with not designing safety aspects into battery operation, as each type has a unique set of vulnerabilities. For instance, as car owners equipped with lead-acid batteries found out to their cost, unless frequently topped up with water to prevent damage by the production of hydrogen and oxygen from the electrolyte when overcharged, the battery would fail. Fortunately, battery technology that's now available to the engineer is considerably more advanced, and battery condition for the sealed units in use today is electronically monitored.

Large-scale energy storage

To meet the demand for large-scale power storage, the Tehachapi Energy Storage Project (TSP) was commissioned by Southern California Edison in 2014. At the time, it was typical of the biggest ***lithium-ion storage*** facility anywhere in the world, capable of supplying 32 megawatt-hours of energy at a rate of 8 MW. Beginning as a combined research project and capability demonstrator, the TSP model was used by Tesla to build a bigger facility in Australia. Constructed right alongside Tesla's Hornsdale Wind Farm, the Power Reserve has a 100 MW capacity and is equipped with Samsung lithium-ion cells providing a total output of 129 megawatt-hours (460 GJ), protecting against load-sharing surges and line blackouts.

The Tehachapi Energy Storage Project in California is one of the world's largest battery storage power stations.

Multiple fuses
To prevent fire or short circuit, ancillary circuits have fuses, which usually consist of an appropriate thickness of wire for the amperage stipulated, contained within a heatproof case that can be plugged in or out of a fuse box. The multiplicity of fuses—several dozen on some cars—constrains a failed circuit without compromising the operation of the entire electrical circuit in the car. By isolating the problem, the vehicle can be driven to a safe place for a fuse change.

Energy storage and relief systems since then have become increasingly common, and the UK has contracts for 200 MW of back-up capacity. Located in Wiltshire, the Minety Battery Energy Storage Project is a part of that and has been equipped by China, with their companies contributing more than 85 per cent of the installation. This is not unusual: electro-mechanical engineering projects are increasingly being owned, managed, and operated by China, which has a distinct hold on the infrastructure resources required for these large projects. But these will be eclipsed in capability by the Manatee Energy Storage Center located at the Southfork Solar Energy Center, Florida, USA, where a 409 MW installation will provide 900 megawatt-hour capacity on lithium-ion technology.

INTERNAL COMBUSTION ENGINE VEHICLES

The standard ***internal combustion engine*** (ICE) needs electrical energy to operate; it is essential for operating the car's starter motor, for providing the electrical spark to operate the internal combustion engine, and for operating all the electrically operated features, from lights to stereo entertainment equipment. Alternators are the means by which motor vehicles produce mechanically driven electrical energy to replace electricity drained from the battery as the car is in operation. The electrical current flows along a single cable from the positive terminal on the battery to the powered component and back through the metal body of the vehicle, which acts as the "earth" for the circuit, to the negative terminal on the battery. This type is known as an ***earth-return circuit***, with current flowing from the positive to the negative.

Most modern vehicles use a 12-volt system. The thickness of the wires providing power to ancillary systems is determined by the function of the device being powered. For example, thick cables deliver power to the starter motor whereas filament wires carry power to the light bulbs, where the energy needed to push the current through a resistance is transformed into heat, glowing white hot in the case of the headlight bulb.

The heavy cable that delivers power from the battery to the starter motor, which produces energy to begin a four-stroke cycle, deliver a spark to the ignition system, and start the process of combustion, is independent of the ancillary, or subsidiary, circuits. Most of these are routed through the ignition switch so that they are switched off when the ignition is deactivated.

There is an increasing tendency for modern cars to be fitted with a wide range of electrically driven ancillary equipment, "selling points" for buyers attracted by increasingly sophisticated systems, such as automatic braking, self-parking, and cruise control. To avoid bundles of wires clustered together in unnecessary confusion, some manufacturers replace wires with ***printed circuits***, usually located behind the instrument panel, copper tracks having been printed on them so that components can be plugged directly in without the need for separate wired connections.

Generating electricity to replace that drawn down by the operation of the vehicle is the job of the alternator, which, as stated earlier, has completely replaced the old dynamo. There are several design choices, but the majority of ordinary domestic vehicles and light trucks usually use the ***claw-pole field construction.*** This design adopts a shaped iron core on the rotor to provide a multi-pole field from a single coil winding, its name derived from the claw shape of the interlocking fingers. With the coil inside, the field current is supplied by slip rings and carbon brushes. Selecting the correct size of alternator when measured on rated power output is important. A safe option is an output equal to at least 1.25 times the total aggregated load likely to be imposed on demand by the collective power drawn down by the main and ancillary circuits of the vehicle in question.

The alternator attached to a four-cylinder Jeep developed by American Motors Corporation and built by Chrysler.

Cutaway of a motor vehicle alternator for demonstration purposes. The claw-pole construction is clearly visible, with alternating wedge-shaped N and S field poles in the center and the stationary armature winding at top and bottom. The engine-driven belt and pulley is visible at the right.

ELECTRIC VEHICLES

While for around 150 years the mechanical development of road vehicles has focused on the fossil fuel-fed reciprocating engine, with electricity vital for its operation and that of the vehicle's ancillary equipment, concurrent with that evolution has been a series of electric vehicles that dispense with the internal combustion engine altogether. The reason why electric vehicles failed to challenge the gasoline engine sooner is down to insufficient technology and the power of hydrocarbon lobby groups. Here, we are only concerned with the technology buried within the design and development associated with different types of electric motor and the power supply they use. However, there is still energetic debate over the type of power supply for these vehicles, and these fall into two types: ***DC brushless vehicles*** or ***induction vehicles***.

DC brushless vehicles

Brushless machines incorporate a rotor with at least two permanent magnets, and this generates a DC magnetic field. This field enters the stator core, fabricated from several stacked laminations, to interact with a current flowing within the windings to produce a torque reaction between the rotor and the stator. Of course, as the rotor rotates, the magnitude and polarity of the stator currents must be continuously varied so that the torque remains constant and the conversion of electrical to mechanical energy is most efficient. Current control is the job of the inverter, which changes from the battery output DC supply to AC.

Induction vehicles

The induction motor was first demonstrated by Nikola Tesla in the late 19th century, when electric cars were considered more likely than cars powered by the internal combustion engine. The stators for this three-phase motor are virtually identical to those for the brushless motor in that they each have three

The Tesla Model S chassis without bodywork or interior fitments, showing the drum-shaped drive motor and inverter between the two wheels, power provided by a set of battery segments laid across the floor of the vehicle.

sets of windings inserted within the stator core. The main difference between brushless and induction motors is the rotor. The induction motor has no magnets, possessing instead a shortened structure in which steel laminations have buried peripheral conductors. The frequency of the magnetic field is sensed by the rotor to be equal to the difference between the applied electrical frequency and the rotation frequency of the rotor itself. This presents an induced voltage at the shortened structure that's proportionate to the difference in speed between the rotor and the electrical frequency. Currents are produced within the rotor conductors that are about proportionate to the voltage, hence the difference in speed. These currents interact with the original magnetic field to produce forces, of which the most valuable is the required rotor torque.

Batteries in electric vehicles

Commonly, a collection of lithium-ion cells provide the battery pack in an EV, and these are arranged in various ways. The cells in a Tesla Model S, for instance, are arranged both in series and in parallel to control the power required. Tesla have chosen to have a very large number of cells arranged in trays across the underfloor of the vehicle, with cooling effected by a glycol solution passed through inner cavities between each battery cell channeling heat to the radiator at the front of the car. It is not uncommon for high-density battery packs to contain 7,000 cells, evenly distributed for good mass balance, low center of gravity, and optimized power drawdown. Early EV designs had only a few battery cells, creating thermal hotspots and a disproportionate mass loading on the vehicle.

The advantages of electric vehicles

All-electric vehicles are mechanically more efficient and considerably more refined than the internal combustion engine (ICE), from the source of mechanical energy through to its delivery. The ICE is not self-starting like the EV and requires several accessories, such as the alternator, to make it work. With an induction motor, the EV has uniform power output and does not require any accessories to work effectively. A typical ICE engine may have a mass of 0.2 US tons (180 kg), a power output of 140 kW and a weight/power ratio of 0.8 kW/kg, while an equivalent EV "engine" will have a mass of 0.04 US tons (31.8 kg), a power output of 270 kW and a weight/power ratio of 8.8 kW/kg. The DC batteries can consist of large or small cell populations according to the type of car design. The DC to AC inverter controls the power frequency and therefore the speed of the motor. The inverter can also control the amplitude of the power and this gives it control over motor output, and in this capacity it can be considered as the "brain" of the EV.

In the Tesla Model S, the drive train incorporates both gearbox and differential and is located in the center position between the electric motor and the inverter. Here, electrical power converted to mechanical energy for operating the drive wheel at the rear is much the same as the differential on an ICE-powered vehicle. Moreover, because the motor is efficient at a wide range of operating speeds, the gearbox needs to be only a single-speed transmission. Selecting reverse is achieved merely by switching the power phase. In the EV, there is also a shift from complex mechanical processes to smart software, with algorithms taking control of respective wheel speed when cornering or turning in a circle.

A unique advantage with the EV is regenerative braking when the power pedal is released, harnessing the kinetic energy of the car in motion instead of dissipating it as heat. This is a process whereby, under braking, the induction motor acts as a generator, putting much more power in the stator coils than the electricity supplied from the battery. The excess power can be stored in the battery pack after conversion from AC to DC. This eliminates the need for more than a single pedal to select and control drive speed, the brake pedal being required only to bring the EV to a complete stop.

The advantages of an EV are many and include high performance, high response rates to control inputs, virtually no motive noise, less pollution from the car in motion, lower maintenance and driving costs, high torque and power output, and highly effective traction control.

When pitted against hydrocarbon vehicles, these plus points stack up, not least their environmental credentials. Many forms of electro-mechanical engineering, especially for hydrocarbon vehicles, are currently being assessed more radically than ever before and governments around the world are paying heed to concerns raised by their electorates about damage to the environment, to the climate, and to adverse weather systems caused by growing levels of carbon dioxide and methane. Unequivocally, hydrocarbon engines are more harmful at the point of use than EV alternatives, if only because of the particulate concentrations and the aerosol mix that accumulates around dense concentrations of ICE vehicles.

However, there are several worrying aspects regarding the moral obligation of the engineer to address issues that are vulnerable to politicization and influences from industrial lobbying. The motor industry is determined to exploit the need for less environmentally damaging vehicles and has seized the EV as the champion for "greener" living as electric vehicles replace the ICE-powered car. Yet all this has been at the behest of politicians, governments, and lobby groups. The "greener" solution was determined to be electric vehicles because the investment required to move away from fossil fuels was less expensive for that class. But while EVs are less environmentally damaging at the curb, in use they are surprisingly harmful, due largely to the lithium-ion batteries they require, the energy required for their production, and the uncertainties about what to do with them after use.

For not only is there a non-existent market for recycling or decommissioning the environmentally harmful and toxic lithium-ion substances, but the way in which lithium batteries are made is also supporting a group of vulnerable workers in developing countries who are employed on substandard wages doing work down mines that endangers their health. Like so many aspects of engineering, a balance needs to be struck between relative levels of damage caused by respective systems in developing a technological solution. Achieving this balance is as much a part of an engineer's work as the mechanical challenges imposed by demanding specifications and requirements.

FUEL CELLS

The fuel cell is another source of energy for the production of electricity and is a suitable alternative to the environmentally and ethically damaging lithium-ion battery used in the EV. Its name erroneously implies some form of combustion but that is not the case. The fuel cell produces electrical power through chemical reactions and not by a heat engine. In its most simplistic form, the fuel cell is the principle of electrolysis in reverse: instead of using an electrical current to split water into its constituent hydrogen and oxygen molecules, those two chemicals are brought together over a catalyst in a pair of ***redux reactions***. In a conventional battery, energy is derived from metals and their ions or oxides, whereas the fuel cell requires a continuous supply of hydrogen and oxygen. While the supply persists, they produce a continuous flow of electrical energy.

Theoretically, there are several candidate types of fuel cell, all of which must have an anode, a cathode, and an electrolyte to move between the two sides of the cell. It is at the anode that the catalyst induces reactions through oxidation to generate the ions and the electrons. As the ions move to the cathode through the electrolyte, the electrons flow in the reverse direction by way of the external circuit, producing direct current. Another catalyst at the cathode causes the ions, electrons, and oxygen to react and to form water with some traces of other products. As we shall see, that was a great advantage for the first truly seminal application of the fuel cell.

Fuel cells are suited to modest power requirements and have an efficiency of up to 60 per cent, although a thermal cogeneration fit, in which the heat produced is

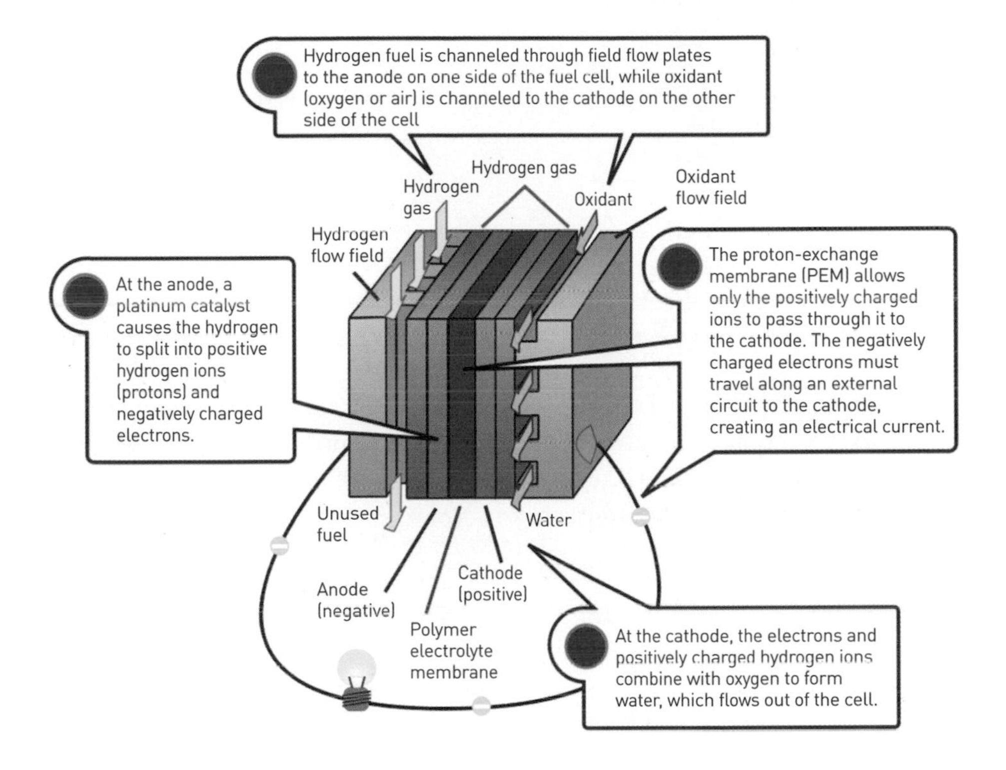

The operating principle of the proton-exchange membrane fuel cell, in which the bipolar plate operates as the electrode with a milled gas channel structure fabricated from conductive composite materials. It incorporates porous carbon papers, a polymer membrane reactive layer, and a polymer membrane.

put to productive use, can raise this to 85 per cent—a phenomenal improvement on the 25 per cent efficiency of an internal combustion engine. The theoretical possibilities for the fuel cell were worked out in 1839 by the Welsh inventor Sir William Robert Grove (1811–96) but the practical application was realized more than a century later. That came about through a range of experiments carried out in Cambridge, England, by engineer Tom Bacon (1904–92) while working at the Marshall Group conducting experimental work supported in the 1950s by the National Research Development Corporation (NRDC).

Known as the ***Bacon fuel cell***, it offered an opportunity for electrical power supply in certain applications for which heavy batteries were inappropriate and solar cells were not sufficiently developed at the time to satisfy requirements. The Bacon fuel cell was first demonstrated on August 25, 1959 to interested parties and the press, producing 6 kW and driving a forklift truck and a circular saw as well as producing power for an arc welding machine. Over the next couple of years, the US propulsion and energy company Pratt & Whitney got involved and from there proposed the Bacon fuel cell for providing power to the Apollo spacecraft.

Fuel cells in space

Apollo had been designed to follow after the tiny Mercury capsule that was capable of carrying just one person for only 36 hours. Apollo, by contrast, was conceived to carry three people and to remain in space for up to 14 days. Clearly, the amount of power required for that duration could easily have been provided by batteries but they would have placed a colossal mass burden on the weight of the spacecraft and required a much heavier rocket than even the giant Saturn V that was used for the Apollo Moon missions. The US space agency NASA wanted a 28 volt DC supply for Apollo with inverters converting to a 115/120-volt, 400 Hz, three-phase power to two AC buses. To this end, three fuel cell power plants were produced by Pratt & Whitney, each comprising 31 cells connected in series. Each cell consisted of separate hydrogen and oxygen compartments and two electrodes, one for each reactant. The electrolyte was a mixture of 72 per cent potassium hydroxide and 28 per cent water, providing a constant conducting path between electrodes. Each fuel cell produced 563–1,420 W or a maximum 2.3 kW.

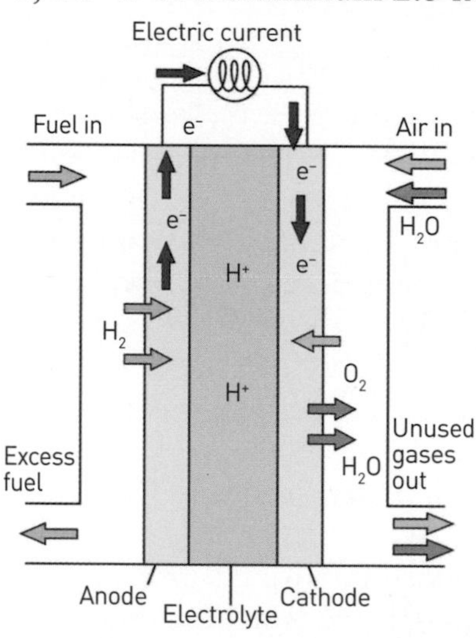

With hydrogen fuel as the input, this is a schematic of a proton-conducting fuel cell as applied to early uses in the space program.

But here it is appropriate to show how a single solution for one problem can be integrated into a weight-saving answer for other design challenges. Clearly, the crew needed oxygen, water for drinking and rehydrating food, and electrical power—all of which were met by the three fuel cells. The oxygen tank that was required to provide a habitable atmosphere for the crew could also be used to supply the fuel cells with one of the two essential reactants; the hydrogen supply had no other application but to supply the second reactant to the fuel cell. And while electrical power was the function of the fuel cells, the water produced by the reaction could be filtered off for use by the crew. By engineering in two separate tanks for each of the two reactants (hydrogen and oxygen) serving the three fuel cells, a wider range of environmental and support solutions were thus provided simultaneously. In this manner, without dedicated systems for power, cabin pressure, and water, a single unifying solution saved weight and reduced complexity.

The first opportunity to test the concept of the fuel cell in space occurred in August 1965 when the two-man Gemini spacecraft ran for eight days in orbit, checking out the system before returning the crew to Earth. There were severe problems on that mission, with the system threatening to break down, but technical solutions were found. Thus, the fuel cell concept was vindicated and the first Apollo spacecraft to carry crew was launched into orbit in October 1968. This spacecraft was designed as the crew carrier for getting to and from lunar orbit, leaving the Lunar Module to take two astronauts down to the surface. The Lunar Module used batteries because the operating life of that spacecraft required support for only two men over a maximum of three days, rather than three men over two weeks, as was the case for the Apollo spacecraft.

Seven Gemini missions and all 15 Apollo spacecraft flown to Earth orbit or the vicinity of the Moon between 1965 and 1975 were powered by fuel cells. Thus, after a flawless performance and 100 per cent reliability on Apollo, that system was selected for the Shuttle. NASA required the Shuttle fuel cells to provide a continuous supply of 21 kW at 28 volts DC and selected three fuel cells each producing 7 kW, or 12 kW at peak output. The fuel cells developed for the Shuttle were considerably more advanced than those built for Apollo but provided the same engineered solution by again utilizing heat output and water production.

The three fuel cell modules for Apollo were installed on a shelf in the service module (the third of which is behind the two visible here), a cylindrical and an unpressurized part of the spacecraft on top of which rested the conical-shaped command module carrying the crew.

The fuel cell used on the Shuttle was a more advanced and evolved development of the concept as applied previously to Gemini and Apollo spacecraft.

FUEL CELLS OF THE FUTURE

The future for the fuel cell is challenging but, as we will see in Chapter 13, there are potential applications that will suggest engineering solutions to environmental challenges. In short, the "hydrogen economy" is every bit as exciting as steam power was in the 19th century and nuclear energy proved to be in the 20th century.

The use of electricity to power homes, industry, and transport has seen the inexorable rise in new methods of power distribution, providing energy for everyday life and exotic functions. But electrical connections can themselves move machines, and radio frequencies are playing an increasing role in the design of a new generation of moving vehicles.

Chapter Twelve

ELECTROMECHANICS

Electromechanical Machines—Controlling Motion—Stealth Technology—Controlling Power—Mechanical Symbiosis

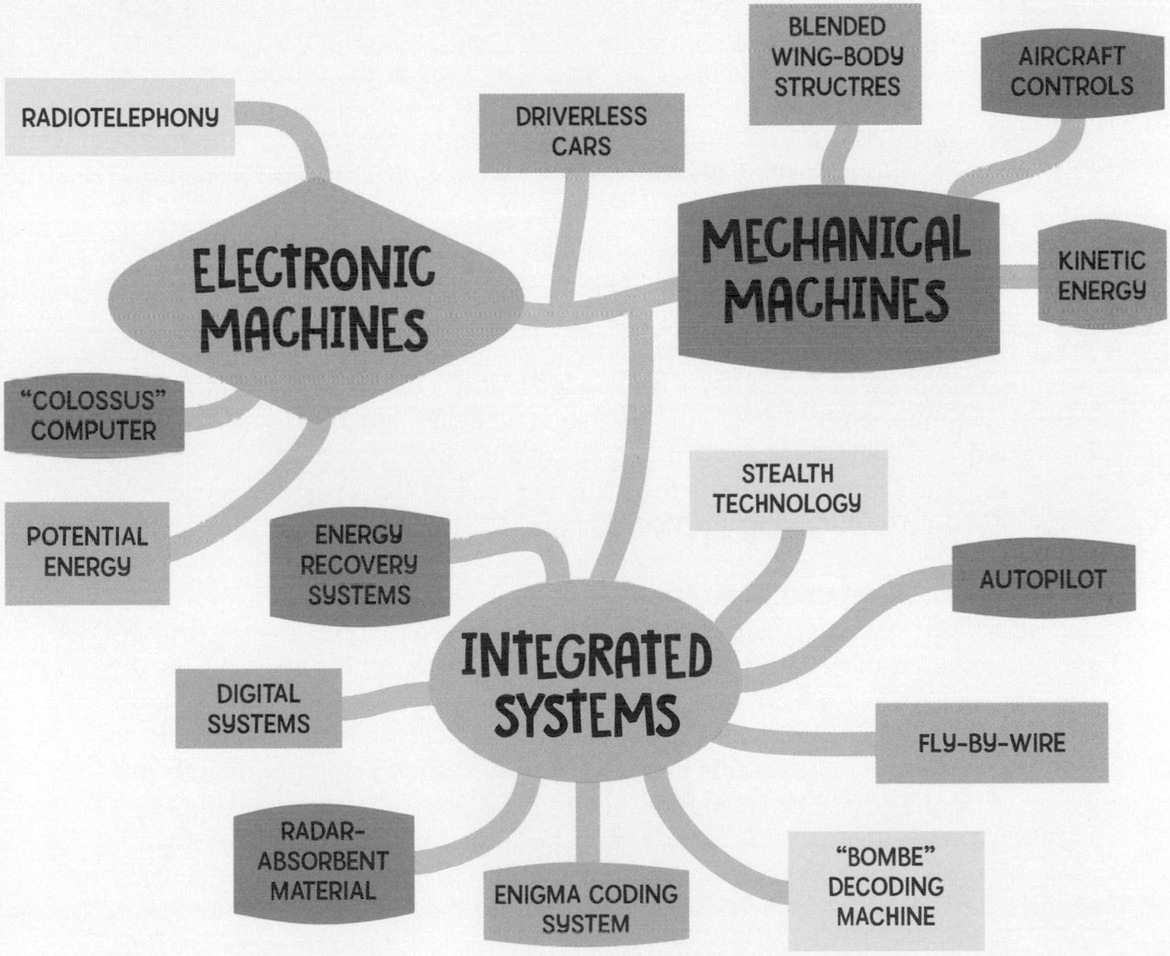

ELECTROMECHANICAL MACHINES

So far, we have identified key agents in the development of mechanical engineering and in electrical engineering. But the unification of the two categories dates back almost to the beginning of the use of electricity to move things. That began when Michael Faraday (1791–1867) put together the world's first electric motor in 1822 and spawned an industry that, while slow to get off the ground, would form a distinct category—***electromechanics***. This category is an applied technology rather than a separate field of research and development and it embraces electrical, electronic, mechanical, and chemical systems. The developed form of electromechanics includes physics, electromagnetism, and combinations of both.

In some respects, electromechanical equipment and its technological application served as a precursor to the field of electronics and while it is still highly relevant today, it was itself the base upon which the modern world of computing machines and artificial intelligence (AI) was founded. It has a long and distinguished history, building both the world of today and potentially that of the future in several diverse ways. In fact, its origins go back almost 200 years.

By the early 1830s, Faraday had demonstrated the use of an electric current to move things, when he passed a magnet through a coil of wire producing an electric current measured by a galvanometer. This was a development of what Faraday had observed much earlier, when he caused a wire that had been placed in a jar of mercury with a magnet at the bottom to spin when the wire was connected to a battery. The magnetic field produced by the magnet interacted with the magnetic field caused by the connection to the battery—a rather elegant demonstration of the principle that the flow of an electric field stimulates a proportional magnetic field.

Electromechanical communication

The inevitable growth of electromechanical devices happened slowly, in part proportional to the application of long-distance communication, specifically the electrical telegraph system in which coded pulses of electrical energy were sent through wires connecting distant places. Today, it is difficult to fully appreciate the impact this invention had when it was first applied in the 1840s. For the first time in human history, the instant transfer of messages between two people did not require that they be within sight of each other. It also instantly obviated the delay inherent in the postal system; it frequently required two months for letters from England to reach China and more than ten weeks to reach Australia. Now, communication was instantaneous, so long as a wire connected the sender and the recipient.

From then on, communications advanced and the use of radiotelephony increased, replacing wires with a wireless technology. By the early 20th century, the use of electrically operated systems to perform mechanical functions presaged the era of semi-automated systems, a significant expansion taking place with the two world wars between 1914 and 1945. It is here that the convergence of electrical and mechanical systems was introduced in great numbers and with broad variety, the invention of the alternator being one example. Moreover, the general expansion of heavy industry during the second half of the 19th century served as a spur to invention and the practical application of devices capable of controlling mechanical equipment.

A letter-printing telegraph set built by Siemens and Halske in Saint Petersburg, Russia, around 1900.

Creating and cracking code

Arguably the first and most elegant demonstration of an electromechanical machine was the Enigma ***encoding cipher machine*** that was used by the German High Command throughout World War II, in various stages of evolution and sophistication. A device made famous by the celebrated operation tasked with cracking the code at Bletchley Park, England during the war, it consisted of a mechanical apparatus designed to vary an electrical current to operate a system of rotors and electrical contacts to "scramble" the message in a way that could be "unscrambled" at the other end by another Enigma machine. Entered in plain text after various plug-board connections had been set up for that particular day, the messages fed into the device would be encrypted and sent in one of the most complex and secure means of communication of its day. It inspired the development of a series of machines, known as "bombes," which became increasingly sophisticated as the war progressed to keep pace with developments with Enigma machines. Further advances with a new generation of encoding machines stimulated what became the world's first highly complex computer in an attempt to decipher the messages. This was known as "Colossus" and will be described in the next chapter.

The Enigma machine was adopted by the German High Command for military communications throughout the World War II. This example, on display at Museo della Scienza e della Tecnologia in Milan, Italy, shows its keyboard, plug-board, and rotor wheels.

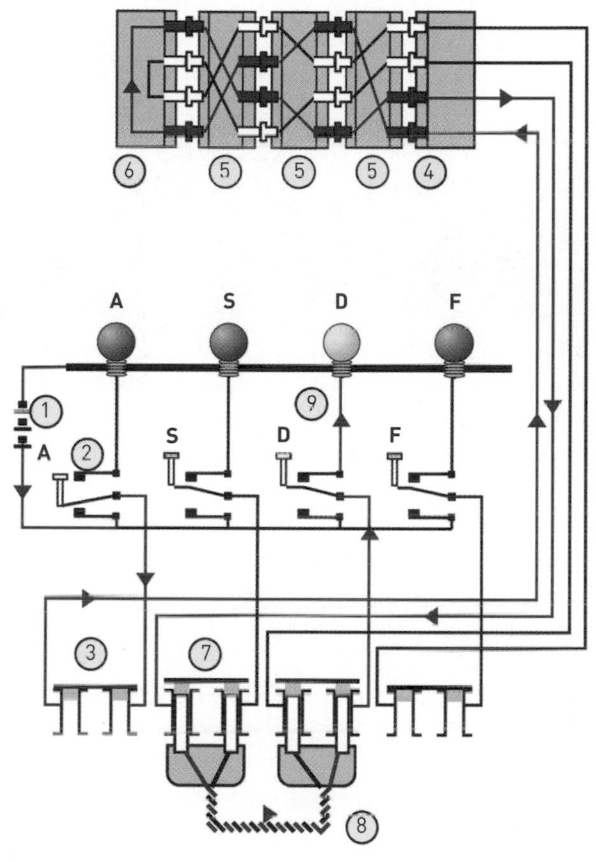

The wiring diagram for an Enigma machine, with arrows and the numbers 1 to 9 showing how current flows from key depression to a lamp being lit. The A key is encoded to the D lamp. D yields A, but A never yields A; this property was due to a patented feature unique to the Enigmas, and could be exploited by cryptanalysts in some situations. In a simplification for clarity, only four components of each are shown, whereas in reality there are 26 lamps, keys, plugs, and wirings inside the rotors. The current flows from the battery (1) through the depressed bi-directional letter switch (2) to the plug-board (3), then the plug-board allows for rewiring the connections between keyboard (2) and fixed entry wheel (4). Next, the current proceeds through the unused and closed plug (3) via the entry wheel (4), through the wirings of the three (army) or four (naval) rotors (5), and enters the reflector (6). The reflector returns the current, via a different path, back through the rotors (5) and entry wheel (4), and proceeds through plug S connected with a cable (8) to plug D, and another bi-directional switch (9) to light up the lamp.

These early indications of the potential inherent in electromechanical systems, while appearing tangential to the main development story in terms of their use today, were vital steps in teaching new generations of engineers to break through the limitations imposed by relatively primitive technology. It would be the development of the transistor and solid-state electronics devices that would prove truly ground-breaking, introducing capabilities only possible through the introduction of sophisticated electronic devices, and these would be stimulated largely by the defense, aviation, and space industries in the second half of the 20th century.

The European manufacturer Airbus became the first mainstream company to integrate fly-by-wire (FBW) into its airliners, the A320 family becoming the first to apply an all-glass cockpit with a digital flight control system replacing analog instruments.

CONTROLLING MOTION

For the last few chapters we have been concerned with mechanical devices used to produce electricity, but this chapter is really about electrical energy used to operate machines and, as such, to move towards computing machines and artificial intelligence. A few examples will clarify how that is put into practice, the first of which is the fly-by-wire (FBW) design concept that is increasingly employed in the aviation and motor industries.

Fly-by-wire in aircraft

The conventional solution to maintaining control of an aircraft in flight is to change the airflow across tab-like surfaces on the trailing edges of the wings (***ailerons***), the horizontal moving tail (***elevators***), and the vertical stabilizer (***rudder***), enabling the pilot to change pitch, roll, and yaw. Historically, connections between the pilot's control column and foot pedals and the actuation of those surfaces were made by cables between mechanical linkages.

Later developments incorporated ***servo motors*** to relieve the pilot of the force required on big aircraft and fast jets. This was a breakthrough, and the first aircraft to be so equipped was the Russian Tupolev ANT-20 during the 1930s. But even so, there was still the mechanical linkage and a direct back-up operating system relying on cables and rods.

The next stage in development—which was a long time coming—arrived with the electronic age in the form of Canada's CF-105 Arrow high-performance combat jet. This was the first equipped with FBW in its purest form, with no other means of control. This aircraft appeared in 1958 but, for reasons completely unrelated to FBW, was canceled. It was therefore not until 1969 that an operational aircraft flew with FBW—the Concorde supersonic airliner. It should be noted that this had been preceded by a range of experimental and semi-operational systems during the 1960s, one sector of this being driven by the demands of the space programme.

Fly-by-wire in space

From the very first US manned flight in 1961, NASA astronauts flew spacecraft that had FBW, first as a back-up and then as the sole means of operating thruster jets to control attitude in the vacuum of space, and other mechanical equipment. Indeed, it was on board both the Apollo and the Lunar Module spacecraft that carried men to the Moon. Moreover, the Lunar Landing Test Vehicle (LLTV), used to give astronauts experience with vertical landings in a device having the control characteristics of the Lunar Module for setting its crew down on the surface, was operated solely by FBW.

The practical advantages were simplicity, reduced weight (by replacing physical linkages and pushrods), and high reliability—because there were almost no mechanical parts, which made for higher ***mean time between failures*** (MTBF). These advantages were countered by the requirement for electrical power—it would be impossible to control an aircraft if it lost power—and by the very human factor of not wanting to relinquish human control to a thin wire! But slowly, and because it made economic sense through reducing the weight of the aircraft, commercial aviation began to look at introducing FBW on passenger-carrying airliners.

As a further extension of FBW and integration with increasingly sophisticated sensors coupled to high-speed processors, multiple standby "redundant" computers now monitor the output of each other and vote according to input—a process that we examined briefly in the last chapter when we discussed the computer-controlled operations of NASA's Shuttle.

Digital systems have played a great part in improving the application of FBW and allowing aircraft to remain within safe operating limits, denying the pilots a manual override if the computers detect that pre-programmed limits are about to be exceeded. This maintains aircraft within pre-set limits constrained by algorithms programmed by the manufacturer. It is necessary because during the early stages of digital FBW and autopilot systems, it was not uncommon in military aircraft for pilots to seize control when they judged the autopilot system was taking the aircraft into a flight path it could not withstand. They could do this because their inputs did not threaten to take the aircraft out of pre-set parameters. However, the intervention could lead to ***pilot-induced oscillations*** (PIOs) whereby the digital computer was commanding the flight controls to operate in one way while the pilot was using FBW to pull it out of that specific maneuver, resulting in a wallowing flight path with the aircraft pitching up and down in rapid cycles. On one occasion when this happened close to the ground, an aircraft slammed into the runway as it went through several vertical oscillations and lost flying speed.

STEALTH TECHNOLOGY

One area where FBW/digital autopilot systems are crafting a new and novel technology (the very definition of electromechanics) is that of stealth technology in military aircraft. Here, the challenges were presented by the burgeoning array of air defense radar systems that could readily detect an aircraft approaching over sea or land. In Chapter 10, we discussed radar and the way it played a seminal role in World War II. That technology has been developed into means by which aircraft can be detected at increasing range. Radar stations placed so that there is no region between their beams where aircraft cannot be detected posed a conundrum: how to reduce the effective radius of the beams transmitted by the radar transmitters. The only way was to make the aircraft less visible on radar, so that they would have to be much closer to the transmitter in order to be detected.

This resulted in what is termed ***low observable*** (LO) technology, a suite of design and materials chosen to make that possible. The most dramatic way to achieve an LO state is to completely redesign the aircraft so that its surfaces reflect the minimum quanta of radio waves directly back to the receiver at the site of the transmitter. Flat planar surfaces on conventional aircraft (large planar surfaces being essential for conventional flight and for making the aircraft stable in the air) are very good at reflecting radio waves. However, by creating a faceted design—in a shape not unlike the many facets of a diamond—radio waves are reflected in many directions away from the radar receiver. What this does is effectively reduce the radius of a radar because so little of the signal gets back to the source of the transmission, minimizing the radius of the transmitter and opening up gaps between fixed radar air defense systems. To these, the incoming aircraft appears like a relatively small bird and does not draw attention, allowing the intruder to slip between the air defense networks and roam at will behind enemy lines.

The search for stealthy technology challenged mechanical engineers to reform the profile of modern combat aircraft into a faceted diamond shape for minimizing reflection of radio waves, as displayed by the Lockheed Martin F-117A.

The measure of an aircraft's visibility to radar is known as the ***radar cross-section*** (RCS), which determines how the aircraft appears on radar. A low RCS means high visibility and vice versa.

For several decades, aircraft design engineers wrestled with this idea but even with digital FBW it was impossible for the pilot to maintain control of an aircraft looking like a faceted diamond, with only the vaguest resemblance to a conventional flying machine. However, an evolution of the digital autopilot, multiple sensors for computer-controlled stabilization and a shape deliberately designed to be inherently unstable could be made to fly if a powerful processor and a complex, mathematically controlled computer program was in charge of keeping the aircraft upright in the air. (Without these design features, it would topple over and crash.) The challenges for electromechanical design in this instance were greater than had been faced before in the arena of combat aircraft design: while the computer held the aircraft stable in pitch, roll, and yaw axes and prevented it from tumbling out of control, the pilot could use FBW to control the aircraft's flight path and maintain authority over where it went, and at what speed and altitude. This "stealthy" design was the pure demonstration of mechanical engineering, with electrical, electronic, and computer-controlled systems fully integrated into the basic design of the aircraft.

What is RAM?
Radar-absorbent material (RAM) literally absorbs radio waves and inhibits reflection. It is applied to the exterior of an object and allows some relaxation of the external shape. It is a key element in securing low observability and has resulted in a variety of different aerodynamic designs.

The first aircraft with these characteristics appeared in the 1980s. Built by Lockheed Martin in the USA, it was officially designated as the F-117A but its pilots knew it as the "Wobbly Goblin" because it was held stable through a series of minute and deft computer-controlled movements, more like a constant set of twitches, from the ailerons and rudders responding to the desire of the aircraft to fall over, and with a continuous series of split-second adjustments indiscernible to the pilot. Attention to detail was unprecedented: even the air intakes for the turbofan engine were covered with grilles so that enemy radar would not get a bright reflection of the compressor blades at the front of the engine. The F-117A proved itself in the Iraq war of 1990–91 when coalition forces evicted the Iraqi army from Kuwait, which they had invaded without warning. First in across enemy lines, it destroyed Iraqi command and control installations together with most of their air defense systems, to open a path for the un-stealthy main force to fly in unimpeded.

Blended concepts

In addition to an inherently unstable platform built around a faceted exterior shape, there was an additional design element that could be utilized for stealthy strike aircraft: ***radar-absorbent material*** (RAM). This revolutionized stealth technology and represented another application of the principles of ***electromechanical engineering***—a blend in which electronic control and mechanical systems come together—albeit in its broadest sense. Various air defense radar systems operate at different frequencies and it was always recognized that the angle of tail surfaces, canted rather than at true vertical, meant that some frequencies would be perpendicular to the incoming signal and reflect back to the source to pin the aircraft as an incoming hostile. RAM overcame this problem.

Iron ball paint

The most common form of RAM is ***iron ball paint***, which consists of microscopic-size spheres with a coating of ***carbonyl iron*** or ferrite. Carbonyl iron has high purity and is produced through the decomposition of purified iron pentacarbonyl, with a chemical formula $Fe(CO)_5$. It takes the form of a gray-colored powder. In its application to iron ball paint for LO technology and a minimal RCS, carbonyl iron is suspended in a two-phase epoxy paint with each microscopic "ball" coated in a silicon dioxide, which serves as an insulator. Immediately upon application and before it dries, the surface finish is given a magnetic field, with the particles held in suspension as the paint dries. Some tests show a preference for applying opposing magnetic polarity on opposite sides of the painted panels, which causes the carbonyl particles to align with the three-dimensional vectors of the magnetic field. It is most effective when the balls are given equally spaced dispersion in the paint, presenting progressively increasing density as viewed by incoming radar waves. On the downside, the science and the chemistry of this is on the margins of practicality for aircraft designed to operate at great speed through the atmosphere and in variable weather conditions involving a wide range of temperatures and pressures.

Did you know?

Carbonyl iron appeared as a commercial product in the 1930s and the research to produce this led to the first magnetic tape. This was eventually replaced with a more suitable substance when applied to conventional tape recorders in common use for recording sound throughout the 1950s and 1960s.

Neoprene polymer sheets

An alternative to iron ball paint is the use of neoprene polymer sheets with carbon particles consisting of 0.3 per cent crystalline graphite, which is deeply embedded in the matrix. These tile sections were used on early models of the F-117A, but they were subsequently removed and replaced with iron ball panels. One of the reasons for this is that the application of the sheets is a highly precise operation that has to be carried out by industrial robots to create the exact thickness and proportional density, which can vary from section to section. In addition, maintenance brings unique demands, which require specialized equipment and highly trained technicians. Both application and sustained reliability therefore call for the very best in electromechanical engineering.

Pure aerodynamism

As a direct product of a combination of the highly computerized flight control systems, stabilization and control electronics, and sophisticated suites of

The blended wing-body planform of the Northrop Grumman B-2 uses cooled exhaust gases to reduce the infrared thermal signature and a serpentine engine intake for engines buried far to the rear to prevent radar reflections off compressor blades.

sensors, a new form of stealth was possible—one in which the purest form of aerodynamics was achievable. Throughout the history of flight, and particularly in the post-World War II era of high-performance combat aircraft, designers have sought ways to eliminate all structure other than the wing, which is essential for lift. Without a fuselage and a tail structure, an aircraft can display minimum drag and high efficiency in the cleanest of aerodynamic shapes, devoid of protuberances that cause disruptive airflow, vortices, and, consequently, ***induced drag***. The reduction of these performance-reducing effects, leaving only ***parasitic drag*** (a combination of form drag and friction drag) from the flying wing, called for major engineering challenges and highly developed electronic flight control systems.

The ***flying wing*** is defined as a blended wing-body structure with no protruding tail surfaces. This means that the bulbous mid-wing position contains the buried crew compartment, perhaps with a weapons bay underneath. This is all that constitutes what would otherwise be the fuselage, which, without a tail surface, is irrelevant anyway. Formerly achieved by the tail, in a flying wing design directional control is effected by adaptation of the ailerons into drag brakes. Unlike a conventional aircraft where the aileron is hinged to move up or down, for the flying wing it is split so that half can hinge up and half can move down simultaneously. This increases drag on that side of the aircraft, which causes that wing to slow and effectively turn the flight line of the aircraft in a yaw maneuver. This elegant solution is made possible only through the computerized, electronic digital FBW installed in the current generation of stealth aircraft.

Eliminating the tail goes a very long way to reducing the RCS of the aircraft and gives it a really low observable signature, further reducing the effective radius of air defense systems and widening the gap through which the flight path of the intruder can be threaded. This step was first made possible by demonstration with the F-117A, which showed that small and compact aircraft designed to be inherently unstable could be combined with a faceted design to create an aircraft with a low observable RCS. But the concept of the flying wing design was better suited to larger aircraft with much greater range, and therefore the requirement to carry more fuel (which dictated their size). The first specifically designed in this way was the very much bigger Northrop Grumman B-2 Spirit bomber, which appeared in the 1980s. Only 21 of them were ever built. The Spirit remains in service today as the ultimate intruder with a "right to roam," only because it can operate almost with impunity behind enemy lines. Nevertheless, the flying wing design is the template from which the B-21 Raider is currently being developed as a cheaper and more advanced B-2 lookalike.

There have been other applications in which stealth technology has been applied, and aerodynamic shaping goes hand in hand with the use of exotic construction methods and advanced frequency-tailored RAM to allow fighters and strike aircraft to assume a more conventional outline. These types include the exotic Lockheed Martin F-22 and the F-35, each of which have varying degrees of stealth for specific roles and applications. All are manifestations of electromechanical engineering in the way they are controlled, but with the design priority to avoid detection through the use of radio frequencies on the electromagnetic spectrum.

The major development of RAM and other engineering solutions has created a generation of aircraft that are capable of re-establishing modern aircraft at the forefront of their traditional role. For a time, it appeared that the role of the piloted aircraft in modern warfare was giving way to automated, unpiloted systems, but the fact that challenges have been met by electromechanical engineering solutions means that a class of technologies have now been created that are crucial to solving less hostile challenges: namely a class of sustainable air transport aircraft—the airliners of the future with totally "green" engines, which are a subject for the next chapter. But existing developments in military aviation will undoubtedly inform the next generation of commercial aircraft, with blended wing-body, possibly tailless, forms reducing demand for fuel and significantly increasing efficiency.

CONTROLLING POWER

The design and development of electromechanical systems requires a degree of modeling to study physical parameters and validate their predictable and safe operation with high reliability. Modeling is itself a highly skilled profession that requires a degree of knowledge of advanced mathematics and physical principles as well as a capability for visualizing the unexpected consequences of a system failure. It is also the place where the mechanical and electrical elements are separated. The mechanical systems are usually modeled around the definition of the difference between the kinetic and the potential energy whereas the electrical systems, while bearing similar comparative evaluation, also relate to power parameters where the equations of motion are specifically defined. From these two, the separate and distinct units converge through a concept around energy within the electromechanical system and in this way issues with either can be resolved via a similarly common set of equations using a sophisticated mathematical process laid down by Lagrange, whom we have met in Chapter 2.

Arguably a sad comment on human progress, much of the stimulus for technological development in the civilian sector originates from the engineering development of military equipment, and this reality equips the engineer for facing new and daunting challenges in the future. In addition, sport—in which engineering plays a dominant role—has its own contribution to make.

In Formula 1 motor racing, for instance, the need for increasing performance, raising efficiency, and significantly cutting carbon dioxide emissions has stimulated progress that is subsequently being applied to motor vehicles used by the general public around the world. In just six years, up to 2019, the average F1 car engine had a 20 per cent increase in power output with a 26 per cent reduction in CO_2 levels. Expressed as a percentage of energy from a combustion cycle, the F1 car achieved a peak level of 40 per cent in thermal efficiency, up from 26 per cent prior to 2014, and now at 50 per cent. That unprecedented increase was made possible by the significant financial

Stealth design concepts have adopted the faceted approach to minimize radar reflections, as displayed by the Swedish Navy's stealth corvette HSwMS Helsingborg, *seen here during a visit to the port after which it is named.*

investment urged on teams by the desire to win, a process in which engineers have made a meaningful contribution to making the conventional motor vehicle less polluting, smaller, and less costly to produce and to operate.

In a secondary sector that involves electromechanical engineering, entirely new concepts involving moving around the energy created in an internal combustion engine have again been developed for F1 racing. The stimulus was a change in regulations, which in 2014 required all F1 racing cars to be turbo-hybrids, and the incentive for that was improved prospects for winning races. The challenge was to design power units of small displacement that would produce very large quantities of energy. In what was known as the MGU-K (Motor Generator Unit-Kinetic), a small electric motor was connected to the internal combustion engine so that it could recover energy from heat that would usually be lost during braking. This heat would be recovered from the kinetic energy and transferred to a battery as potential energy. In an F1 car, that motor powered by the recovered energy is mechanically connected to the crankshaft through the timing gears, and helps produce more energy for transmission to the rear wheels.

Another form of energy recovery is the MGU-H (Motor Generator Unit-Heat), which takes energy from the turbocharger, stores it and uses it to spin up the compressor at the other end. The exhausted gases from the engine cycle spin the turbine, which in turn drives the compressor, but since the MGU-H is positioned between the two, in addition to spinning the turbine it produces electricity, which is then stored in the battery. As the car accelerates, the electricity spins the compressor to provide an immediate power boost rather than waiting for the turbine to spin up following a brief lag. The technology has been applied to e-turbo road cars, allowing small engines to produce high power output, reducing the fuel they burn to achieve that level of output as well as cutting toxic and environmentally damaging emissions.

It is this technology that has produced the ***hybrid electric vehicle*** (HEV) in all its many and varied forms. ***Parallel hybrids*** are those that have both an internal combustion engine and an electric motor connected to the mechanical transmission, providing the power to drive the wheels simultaneously. The power output of the electric motor has increased to well over 50 kW in current vehicles. The balance of power between the two motive forces can produce a highly efficient powertrain but there are other options for different requirements and specifications, largely driven by market demand and the requirements of the owner. A ***series-hybrid*** design places the electric motor as the prime source of motive power while a small internal combustion engine is there to provide additional power to the battery. This is particularly useful

A BMW Sauber V8 engine for the 2006 season represented an intermediate stage in the use of innovative engineering for F1 cars.

for extending the range of the vehicle so that the electric motor has a life-extension capacity and is also beneficial in urban environments where congestion or on-board power demand (such as from lights or a heater) can readily exhaust the battery capacity. In more sophisticated examples of powertrain options, a ***power-split hybrid*** combines the technology and the efficiency of both parallel and series-split hybrids, and this produces a more efficient balance between all performance demands.

In all of these examples, regenerative braking can be used to convert the kinetic energy from braking into a form that can be used for immediate motive advantage or stored in a battery for use when required. This is another technology directly applied from the F1 industry to the production motor vehicle for everyday use. The heat generated by friction between the brake disc and the brake drum is therefore redirected to a productive force. This is a spectacularly successful combination of integrated energy direction, in much the same way that, as we saw in Chapter 11, fuel cells in Apollo spacecraft and the Shuttle could be used to provide electrical power and water for drinking, with a common supply of oxygen as one reactant and a supply of hydrogen as the other. This is integrated systems management using both mechanical and electrical interchange.

MECHANICAL SYMBIOSIS

In this chapter, we have seen how the combination of electrical control over mechanical functions can symbiotically work to the advantage of both as each exchanges potential or kinetic energy with the other. This is a technology that found expression through the demands of the military for stealthy combat aircraft and within the motor racing world for more integrated and energy-efficient systems involving the conventional internal combustion engine, electric motors, and regenerative braking. All of this can be applied to the challenges faced by demands for motor vehicles which are both more efficient and contain mechanisms that will raise environmental standards for a more sustainable basis on which to build a future world, devoid of polluting chemicals.

Chapter Thirteen

COMPUTING MACHINES AND ARTIFICIAL INTELLIGENCE

Engineering the Future—Automobiles and Aerospace—Small is Big—Biomechanics—Power Production/Distribution—Integration of Engineering Systems—Roles/Responsibilities

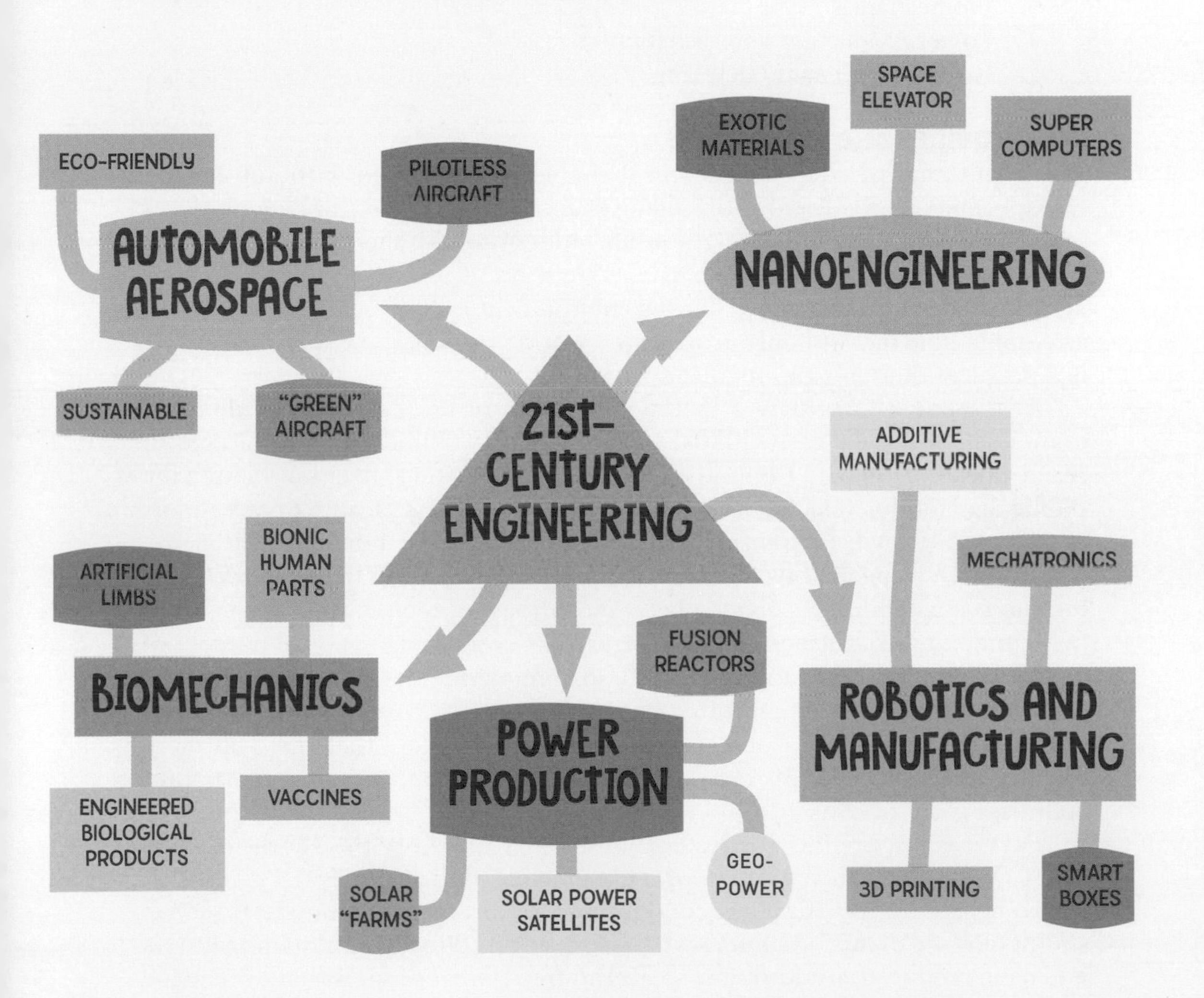

ENGINEERING THE FUTURE

Excluding the ever-present role of the civil engineer in creating buildings of the future and urban spaces supporting a healthy and satisfying life, there are key indicators about the future of engineering in this century and beyond. These can be broken down into several separate categories, none of which existed at the dawn of the modern industrial age:

1. Automobiles and aerospace.
2. Nanoengineering and associated technology.
3. Biomechanics.
4. Power production and distribution.
5. Robotics and manufacturing.

AUTOMOBILES AND AEROSPACE

The depth and breadth of mechanical engineering involved with automotive transport has been covered in preceding chapters, so here we will look briefly at specifics. Like so many challenges in the modern world, there is an overriding need to find more economically sustainable ways of providing power to move vehicles. The damaging effects on the environment of hydrocarbons employed as combustible fuel will only grow as the world's population expands and that is a problem that only the engineer can solve.

At the end of World War II, the population of the Earth was little more than 2 billion. In the early 2020s, it has reached 8 billion and is projected to reach a peak of almost 11 billion by the end of the century. In extrapolating the public demand for vehicles and air travel, the engineer is faced with an almost exponential growth in transportation as measured per individual/kilometre. This is the challenge involved with effecting a complete move from hydrocarbon fuels to sustainable fuels, divided between various technological solutions. The magnitude of the challenge can be measured by comparing the total number of vehicles on roads today and the uptake in coming decades. While it has taken more than 120 years of motoring to grow the global use of motor vehicles to 1.2 billion (in 2020), conservative projections indicate that this will more than double in the next 30 years as rich countries expand demand and poorer countries grow richer.

For air transport, the situation could show greater growth over the next 30 years. While experiencing a temporary hiatus due to the COVID-19 pandemic, which began in early 2020, the total number of aircraft in commercial service stands at more than 23,000. It is estimated that by 2040 aircraft manufacturers will have produced an additional 39,300 aircraft. Great progress has been made in achieving unprecedented levels of efficiency, significantly reducing toxic emissions from aero engines, drastically cutting noise levels around airports, and reducing the number of people required to maintain and operate them. This has a beneficial effect on airport ground traffic and localized pollution from road vehicles. Where once the mechanical engineer supplied bigger and faster aircraft, today his and her skills are focused on reducing carbon emissions, lowering associated pollution levels, and making the aircraft and airports still quieter. One way of achieving that, say some engineers, is through the airliner powered by electric motors, eliminating the need for hydrocarbon engines.

Rolls-Royce engineers preparing an UltraFan variant of the Trent aero engine that promises an efficiency improvement of 25 per cent over the first-generation engines of this type, delivering significant weight, noise, and fuel burn reductions.

NASA's Pathfinder project involved two aircraft for studying solar- and fuel cell-powered propulsion systems, technologies that may one day result in flight free from hydrocarbon fuels, in addition to being quieter and more efficient.

Did you know?
Some aircraft have circumnavigated the globe by obtaining power from photovoltaic cells and lightweight batteries driving propellers, reaching extreme altitude and gradually losing height on the night side of the Earth before recharging in sunlight and climbing once again.

Aircraft with propellers driven by electric motors have been around for a very long time and are much more common than is generally appreciated. Aircraft need motive power to move forwards so that air flowing across an aerofoil can generate lift. Many attempts have been made, some successfully, to provide that power from batteries or from solar cells placed on the upper surface of the wing. In some cases, ***unmanned aerial vehicles*** (UAVs) powered by fuel cells have been equally successful, and several companies have formed to produce these aircraft.

Potential applications have stimulated engineers to focus in two directions: lightweight materials and high-power production/conversion properties. In turn, these have been applied to more conventional aircraft and have stimulated research organizations, including NASA, to focus on this as a potential design template for future commercial, passenger-carrying aircraft.

Manufacturers, including Airbus and Boeing, are funding prototype concept designs in preparation for the day when electric aircraft become practical. But there are many challenges faced by the mechanical engineer when it comes to dealing with the problem of commercial aircraft. These fly high and fast and have an inherent take-off weight between 66 tons for a small regional airliner such as the Boeing 737, accommodating around 150 passengers, and 634 tons for the long-range, high-capacity Airbus A380. In theory, this mega-airliner can carry 850 people but most commonly, with its present seating arrangement, about 550. Typically, 43 per cent of an airliner's maximum take-off weight is in its empty weight, an additional 33 per cent is given over to jet fuel, and 24 per cent is accounted for in passengers, baggage, and cargo. There are various ways to measure fuel efficiency, and these are affected by the size and capacity of the aircraft, the routes flown, the weather conditions encountered, and the overall efficiency of the engines.

Since jet air transport began in the 1950s, engine efficiency has doubled; propeller-driven airliners of the period were up to 28 per cent more energy intensive than modern jet airliners, which fly up to 80 per cent faster, thereby spending less time in the air on a given route for a standard passenger/cargo load. There is, however, a sting in the tail: if the engineer is faced with reducing overall carbon emissions from the use of fossil fuels, the high growth in passenger flights greatly outweighs the 55 per cent improvement in engine efficiency, as measured per passenger/liter of fuel consumed. Therefore, while there is great progress in making engines "greener" and aircraft more fuel-efficient, the increase in demand for air travel is undermining the overall global objective of reducing carbon emissions. Moreover, there are disturbances to the upper atmosphere caused by the production of water-vapor condensation and other undesirable modifications brought about by molecular modification of particulates emitted by the operation of the aircraft, in addition to that produced by the engines alone.

For the engineer, the challenge shifts according to where the parameters fall: merely measuring fuel efficiency relates only to the economic projections of the operator and will always figure highly when it comes to deciding which specific aircraft to buy, whereas measuring the overall impact of air travel on the environment could give the engineer a bigger challenge, because the tourism industry and the surge in low-cost travel in addition to frequent-flyer programs has already caused a sharp increase in demand for air over surface transportation. Engineers are therefore frequently required to provide data that may not be in the preferred interests of all parts of the industry engaged in promoting and selling seats on airliners. One example is the recognized calibration of distance on air or ground transportation, whereby operators acknowledge that the crossover between each is around 370 miles (600 km). Distances less than that are less harmful to the environment if they are taken by surface routes and can be completed in less time when evaluated door-to-door, more so in an environment where check-in time at airports frequently takes longer than the flight itself.

With ever-pressing urgency, the engineer is required to produce machines that can cut costs for the operator, and increasingly this is being accomplished on a scale unprecedented in the history of manufacturing and distribution. Paradoxically, the larger the requirement in terms of scale, the greater the demand on the engineer to use ***nanotechnology***. And this includes several minor changes to engines and airframes—modifications that can be introduced quickly and at low cost. Typical are the upturned wing tips, which provide

Boeing's 2008 fuel cell demonstrator, based on a Diamond HK36 Super Dimona glider.

about a 2 per cent increase in aerodynamic performance, translating into fuel economy. But, increasingly, the engineer is constrained by the economic imperatives of the operators. ***Biofuels*** for aircraft are a classic example of that. Engineers have introduced changes to engines that would allow large-scale use of fuels from renewable resources such as algae, wood, agricultural waste, and some plants, yet there has been very little uptake, despite the US military experimenting with them and despite the 80 per cent reduction in an aircraft's environmental footprint that the use of biofuels offers.

Other technical improvements to performance could also significantly improve the environmental performance of aircraft; ***open-fan engines*** are one such improvement. These resemble conventional turbofan engines but with the frontal shroud removed so that it has the appearance of a turboprop aircraft but which drastically cuts fuel burn and improves efficiency. There are challenges with this. The cowl that covers the fan blades also serves as a protection in the event that a blade should come loose and slice into the pressurized cabin containing the passengers. (Such accidents have happened with cowled engines.) But this is not an insoluble problem and the reason open-fan engines have not been introduced is not for safety reasons, rather a problem implicit within the decision structure at corporations and major organizations. Without political or real public pressure, to an extent the mechanical engineer can only propose changes that are better but that require investment or a complete change in the support infrastructure, which, in itself, can cost the savings made over several years of operations.

An experimental prop-fan engine installed on a Yak-42E-LL airliner, at the 1991 Paris Air Show. Interest in this type of engine, offering the performance of a turbofan and the economy of a turboprop, has been strong in recent years.

SMALL IS BIG

The second category of defined areas for priority in mechanical engineering relates to ***nanoengineering*** and involves everything related to an object measured in scale of 10^{-9} m. This is invariably the field of ***materials research*** and ***biomedicine***. It is where new materials with exotic properties are defined, and rarely strays very far from the world of chemistry rather than physics. And there are some outstanding concepts with real-world potential of such extraordinary possibilities that for the very present still occupy the realm of science fiction. One such example is the ***space elevator***, a planet-to-space transportation system involving a single thread of extraordinarily light weight and strength that could be made to connect a point on the Earth with a satellite in space. This is not so outrageous an idea. NASA is already hard at work studying the potential; in theory it could work well but in practice it remains a future possibility—yet a highly respected possibility.

BIOMECHANICS

Back on Earth, ***biomechanical engineering*** is used to address the needs of amputees and those with diminished physical capacity. It concerns the engineering and satisfactory application of bionic elements to replace missing limbs or hands and feet. Here, too, nanoengineeing is key and the seamless integration of multidisciplinary zones of research and development help knit together capabilities. Biomechanical engineering thus speaks to the needs of humans at their most basic level and provides a window of capabilities that, in only the very recent past, was itself the stuff of science fiction. Yet, the ability to design and manufacture artificial

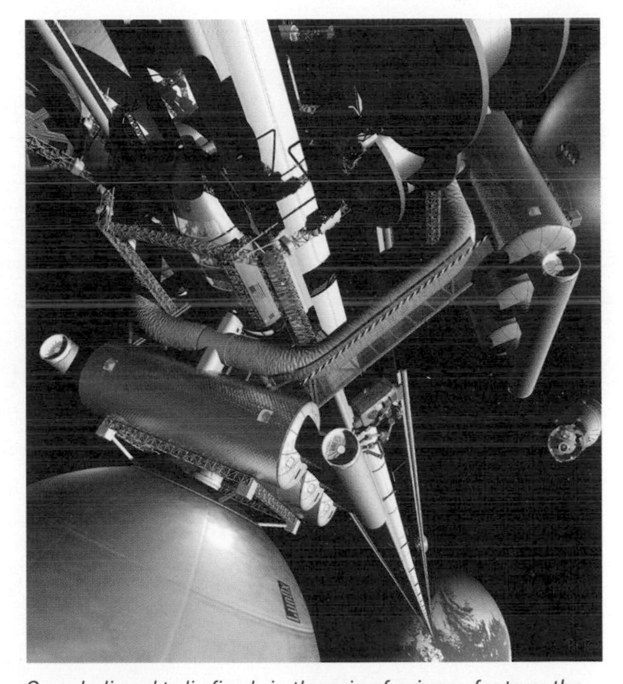

Once believed to lie firmly in the grip of science fantasy, the space-elevator concept—whereby a single structure links the surface of the Earth and a satellite in space—has moved now from science fiction to a potential reality.

Energy inside the Earth
Geothermal energy refers to the natural energy sources stored inside the Earth, and is produced by radioactive energy decay and from the continual heat loss from the formation of the planet 4.5 billion years ago. Tapping into this source is highly cost effective and sustainable, and is ecologically friendly to the environment. It operates on the principle that heat allows productive work to be obtained through its use to drive turbines and produce electrical energy without burning hydrocarbon fuels.

appendages to the human body and have them operate through nerve impulses—just like a naturally grown element from the foetal stage—is just one pioneering area showing tremendous results and outstanding success. And despite the fact that it relates to a living organism, it is engineering, every bit as much as the engineering that is applicable to the development of new and sustainable transportation systems.

POWER PRODUCTION AND DISTRIBUTION

The search for sustainable power production for distribution to national grids is supporting a wide range of potential engineering solutions and, while being on the margin of mechanical engineering, it constitutes a field of research and development that could have profound implications. Some of these are purely technological applications, and include a research project to develop ***betavoltaics*** in which nuclear waste would be used to produce electricity. It works by using a waste product, strontium-90, as a conversion device, employing the two high-energy electrons emitted during its decay to achieve efficiencies of up to 18 per cent and develop 1 watt of energy continuously for 30 years, which is about 40,000 times the current capacity of lithium-ion batteries.

The delivery of electrical energy for the use of industrial, urban, and rural communities has challenged engineering for more than a century, and some extravagant alternatives to the use of fossil and nuclear fuel to drive the production side have been proposed in the last several decades. Some proposals are highly sophisticated and probably best left to a future century, but two are noteworthy for their engineering challenges: ***solar power satellites*** (SPS) and ***nuclear fusion***. Each shows practical possibilities, but the technology and the investment required to achieve them are at different levels to most engineering solutions.

Solar power satellites

The SPS is an idea that was brought to serious analytical study during the global oil crisis of the 1970s. It involves the assembly in geostationary orbit of large arrays of solar cells, each occupying several square miles in area.

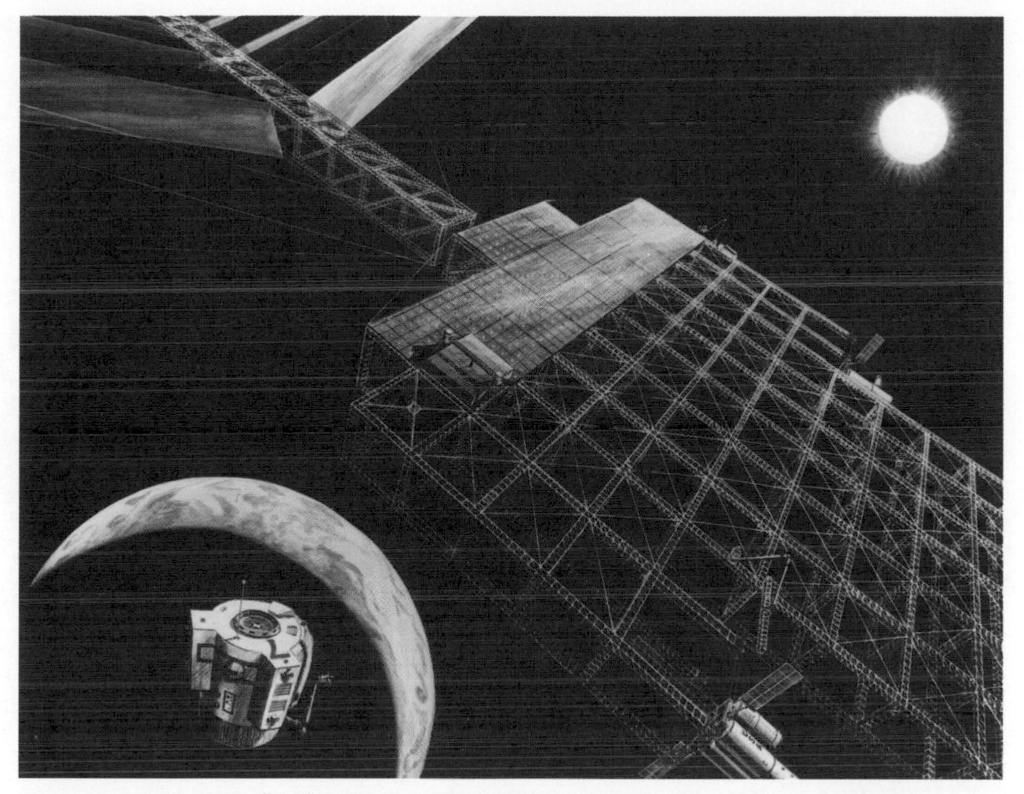

The solar power satellite (SPS) concept of 1976 appeared to be a way of providing increased energy to satisfy a global explosion in population levels. It envisaged very large arrays of photovoltaic cells to produce electricity that would be beamed down to Earth on microwave links and converted into current by diodes in rectennas on the ground.

Through photovoltaic conversion of solar radiation into electrical current, the SPS would transmit power as a microwave beam to ***rectennas*** on the ground. For such an ambitious project, each rectenna would be equipped with several diodes. The SPS became the largest funded study in a joint program between NASA and the US Department of Energy. It was proven to be feasible and a program was proposed that would replace every ground-based power station on Earth with a network of such solar "farms" in space, spread around the equator in geostationary orbit 22,245 miles (35,800 km) above Earth. However, the limiting factor was cost and the sheer logistical size of launching so much mass into orbit, assembling it, and servicing the elements—both in space and on the ground. With the rise of a political solution to the oil crisis plus a new administration, the 1980s saw a decline in interest.

As an alternative to the SPS concept, NASA and the US Department of Energy examined the possibility of collecting solar energy and beaming that down to Earth in giant power stations in geostationary orbit, which, like the SPS, would appear to remain over a fixed spot on Earth.

Nuclear fusion power

Nuclear fusion power is another potential solution to the use of unsustainable and limited resources for producing electricity but, like hydrothermal and fossil-fuel-fed reactors, the ultimate objective is to provide energy for spinning a generator. ***Fissioning*** of the atomic nucleus of a heavy element has been discussed in Chapter 9, where the production of electricity and the development of an atomic bomb were developed together. The fission process provides for the liberation of infinitely greater quanta of energy by fusing light atomic nuclei, and the first demonstration of that on a large scale occurred on November 1, 1952 during the test by the USA of the first ***thermonuclear reaction***. For engineers, the big challenge has been to harness fusion energy within a small volume in a nuclear reactor and use it for generating electricity. If this can be achieved, it will provide the true promise of "clean" nuclear energy without the waste products that are created in the fission reactor.

The design of a fission reactor is relatively simple: the fissile fuel will trigger a sustained chain reaction if sufficient fissionable material is brought together in a pile. The heat produced needs be no greater than the thermal transfer of energy required to change water into steam for driving the turbine to spin the generator. The fusion process is a demanding challenge for engineers because the core in the reactor has to be hot enough to ensure adequate thermal distribution to overcome the repulsive force between the light atomic nuclei, known as the ***Coulomb effect***. The force of repulsion increases on the inverse square of the distance separating (for instance) an electron and a proton and this is about 10^{10} relative to the force within the nucleus between those two particles. The Coulomb effect is counterbalanced by the strong nuclear force between the nucleons (protons and neutrons) and has the strength within its range to dominate the effect of repulsion.

Fusion of light nuclei in the Sun takes place at a temperature of approximately 1 kiloelectron volt (keV), which is about 10.6×10^{6}°C, and from the fusion of two protons produces deuterium with a positron and a neutrino. Because the Sun is so large, it can sustain fusion from a low-probability event for a considerable length of time; it takes 10^5 to 10^6 years for energy from the leaking core to reach the surface. The problem for the engineer building a fusion reactor on Earth is that the reactor core will have a diameter in the order of several metres rather than the 1.4×10^6 km diameter of the Sun. Our home-built reactor would have energy confinement times measured in seconds. Moreover, it would require much higher operating temperatures, which, in the preferred method of magnetic confinement, requires temperatures of 20 to 40 keV. This is akin to the temperature at the accretion ring of a black hole, and is found almost nowhere else in the universe. Production of such temperatures was quite common in fusion experiments during the 1990s but in seeking to reach the break-even point, where fusion energy released exceeds the applied heating power, engineers were unable to come anywhere near the required confinement time for a practical design.

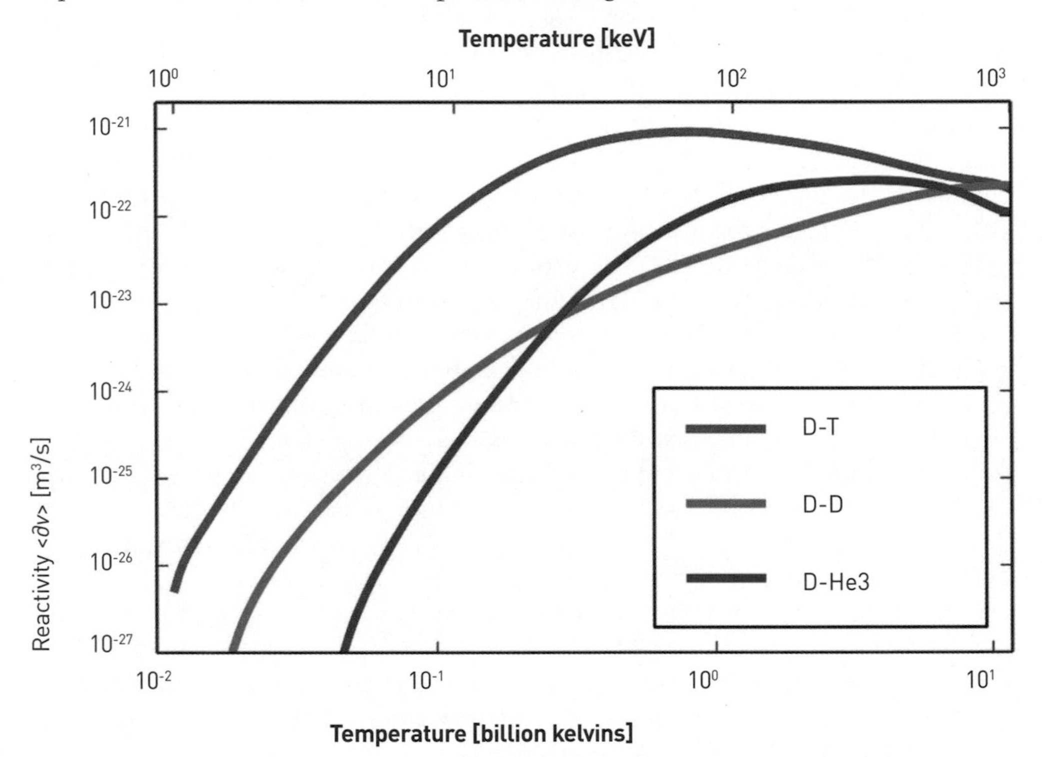

A plot of fusion reactivity (the average of the cross-section times the relative speed of reacting nuclei) versus temperature for three common reactions. The fusion reaction rate increases rapidly with temperature until it maximizes and then gradually drops off. The deuterium-tritium fusion rate peaks at a lower temperature (about 70 keV, or 800,000 K) and at a higher value than other reactions commonly considered for fusion energy.

There are also other technical difficulties relating to the energy spectrum and birth environment of neutrons, and to the colossal challenge with thermal extraction: in a fission reactor, all the energy is absorbed in fuel pellets and distributed throughout the core; in the fusion reactor, 80 per cent of the energy is deposited across a distance of about 3.2 ft (1 m), but the remaining

20 per cent consists of helium nuclei (alpha particles) deposited in the reacting medium, and these have to be extracted on plasma components at high energy density.

Arguably, the greatest challenge for the nuclear engineer of the 21st century is containing the tritium produced from deuterium in the fusion process. This decays to ^{3}He through beta emission and burns to form water, which can be assimilated into biological systems, with the added complexity of it permeating a range of materials and substances. A typical fusion plant of the future could have in excess of 22 lb (10 kg) of tritium, and a great amount of effort has been expended to minimize the quantity produced, with some satisfactory results. There is also a need to contain the production of carbon-14 (^{14}C), which has a half-life of 5,730 years and can be incorporated into DNA with damaging effects. But these are not insoluble problems and are far less injurious when taken in the round than the waste from conventional fission reactors, which have been producing electricity since the 1950s.

Located at Culham in Oxfordshire, UK, the Joint European Torus (JET) was part of a pan-European research project into nuclear fusion for producing grid electricity. At the heart is a doughnut-shaped vessel 20 ft (6 m) across and 7.9 ft (2.4 m) high, obscured by the myriad heating, cooling, and measuring systems that surround it. The large orange limbs are iron, for concentrating the magnetic field that controls the hot gases inside the vessel at temperatures up to 360,032°F (200,000°C). A tower houses the eight neutral beam heaters, which use 100,000 volts to shoot gas into the vessel.

INTEGRATION OF ENGINEERING SYSTEMS

Our final category is that which involves both electromechanical and mechanical engineering and supports the increasing demand for "***intelligent*** ***machine systems***. These are initially programmed to operate autonomously on algorithms written by their designers or to work in synergy with other electromechanical systems and to make independent decisions.

The road to where we are today with computerized systems began several decades ago and can be placed at a variety of different trigger points. Arguably, it may be located at the decision in the UK to build an electromechanical computer called "Colossus" to assist with deciphering complex encrypted codes used by the German High Command during the later years of World War II. Built to decipher the German Lorenz system, Colossus was designed to use a form of Boolean logic (see Chapter 8) for counting operations employing thermionic valves in this pre-transistor period. In a completely different development to that which employed "bombe" machines at Bletchley Park, developed by Alan Turing (1912–54), Colossus was conceived in a section headed by mathematician Max Newman (1897–1984).

In 1994, a team led by Tony Sale (right) set about the task of replicating the Colossus decrypting machine at Bletchley Park, successfully reconstructing its operation and demonstrating the seminal role it played in the history of electromechanical computing.

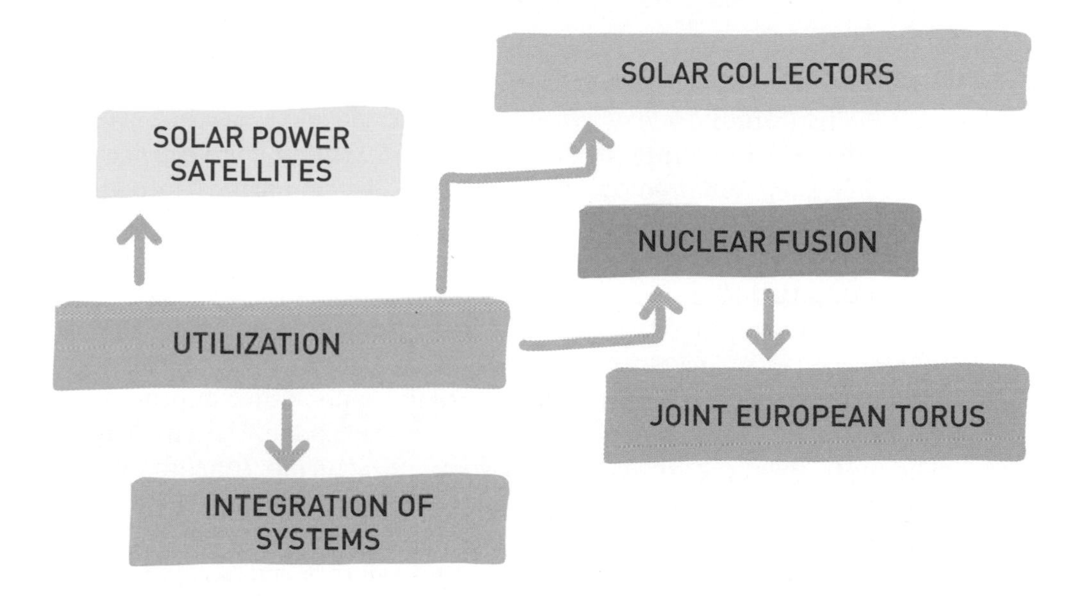

An additional team, led by Tommy Flowers (1905–98), developed electronics and the ***photoelectric reading system***. This had greater security classification than Turing's device and was ready for use by early 1944. Because it was an electromechanical machine, there were problems with the synchronization of two parallel tape loops, but it nevertheless stands as the origin of a series of developments that can be traced right down to solid-state electronics and the digital revolution. As such, at the time it was recognized as a unique development and its design details and two machines were taken to GCHQ (Government Communications HQ), then at Eastcote and now at Cheltenham.

The further development of machines of this type evolved rapidly into more complex and capable devices that underpinned the rise in telecommunications through the two decades following the war. The two fields embracing national security and commercial applications stimulated an ever-widening field of specialized markets, on the one hand leading to highly sophisticated Earth-orbiting satellites for gathering intelligence and reconnaissance information in several bands of the electromagnetic spectrum; and on the other hand, an ever-increasing array of electronic devices for commercial use.

During the 1960s, with the dawn of the Space Age, greater demand for the autonomous operation of electromechanical equipment equalled and in some ways exceeded defense-related, state-of-the-art demands from science and engineering. Where once the military had been at the cutting edge of highly advanced engineered systems, the engine behind new technological developments shifted to the space program and consumer goods of ever-increasing sophistication. Today, there are several niche sectors, which include ***additive manufacturing***, ***mechatronics***, ***automation engineering***, and ***biomechanical design and innovation***, which we will examine in slightly more detail now (apart from the latter, which we discussed in the previous chapter).

Additive manufacturing

Additive manufacturing, or 3D printing as it is also known, involves adding layers of material in a succession of passes over a given base to construct a three-dimensional object, inverting the usual manufacturing process of stripping away material from a solid block to form a shape, thereby leaving waste. The most advanced forms of additive manufacturing are employed in exotic products, such as components for rocket motors and motor vehicles. The process relies on digitally produced computer-aided design (CAD) software used to define separate sections of a specified object in thin layers, allowing the print head to build up the object through a nozzle dispensing the base material. Some machines, by contrast, dispense a powder in a series of many layers, using a laser or electron beam to then melt and fuse the particles together. This not only eliminates the waste involved in cutting down rather than building up a product, but it also eliminates the need for dies and moulds.

NASA astronaut Barry Wilmore holds a 3D-printed ratchet wrench that was produced in space by an additive manufacturing machine on the International Space Station.

Automated manufacturing and the robotic assembly of components into road vehicles has long been a part of factory assembly lines, but a new variation is increasing the sophistication of choice to the level where customers can virtually "design" their own vehicle according to accessories, color schemes, and fittings.

Mechatronics and automation engineering

Mechatronics and automation are outgrowths of electromechanical engineering and include robotics, electronics, and computer sciences together with control engineering in an integrated whole that puts the economic use of time, investment, and development central to how a requirement is met. On several fronts, this is the future of electromechanical engineering. It can be viewed as a stepping stone to automation engineering, where physical devices are controlled by complex electronic control mechanisms. It takes the concept of systems engineering to a new level, where every element of design, assembly, testing, development, and implementation are fully integrated into a single, mathematically defined tool. This is made possible through the connected net-centric web linking digital information processing with programmable robotic devices carrying out labor-intensive functions and standardized work routines.

Quite mistakenly referred to as "artificial intelligence," the most advanced mechatronic and automated devices available today are no more than programmed machines and elements operating on advanced and highly sophisticated algorithms. They are no more capable of deterministic thinking than are the materials from which they are fabricated, and their categorization as such detracts from the true value of such a system—which is in the way an integrated system can operate co-operatively with other machines to improve efficiency or open new possibilities. One such example is ***smart-box engineering***, which elevates mechatronics as the successor to what we have been calling electromechanical engineering. Smart-box applications have emerged from demands by mass-retail companies for automated handling, sorting and the needs of dispatchers to improve efficiency, reduce costs and increase productivity in response to expanding demand. This is all driven by the retail sector and is financed by the savings and potential for expansion inherent in smart-box operations.

The next step for mechatronic engineering is to have additive manufacturing connected to the smart-box "internet of things" so that no human hand touches the process from raw material to arrival at the doorstep of a customer, in either a dense urban environment or to a remote island location where delivery would be through a connected system of ***unmanned aerial vehicles*** (UAVs). With developments in automotive engineering promising autonomously driven vehicles, smart-box technology could effectively supply most customers with a seamless train of delivery chains, each link of which is automated and integrated into a single operational strand.

In other fields, mechatronics can update existing systems in Earth-orbit, space-based assets that are essential to everyday living, replacing separately launched infrastructure products, such as telecommunication and navigation satellites, with remotely manufactured and autonomously launched replenishment systems.

Navigation satellites and interconnected telecommunications worldwide support the mechatronic world of the future, where manufacturing and transport systems (employing autonomous vehicles—perhaps even ships and aircraft) can be controlled by one person operating a single laptop. In such a world, goods sourced halfway around the

Automation connected to sophisticated sensors and algorithms is fusing technology for the new age of mechatronics. With passing gaps of $^1/_5$ in (5 mm) and speeds of up to 13 ft/sec (4 m/sec), smart-box containers harvest products from stowage bins coded and identified by sensors that tag each item, filling a retail order for automatic delivery to the customer.

globe would be delivered via sustainable surface transport and manufactured products could be produced using environmentally benign processes powered by pollution-free energy grids.

ROLES AND RESPONSIBILITIES OF THE ENGINEER

The role of the mechanical engineer has shifted towards a multi-disciplinary function embracing microtechnology and nanoengineering right down to the atomic level. The future for engineering lies along a very different path to that which has prevailed for the last 200 years. Where once it was possible to achieve great feats of engineering "because we could," today we live in a world where the consequences of any major project causes a significant environmental impact on the world. Within the last 100 years, there have been increasing concerns about the environmental damage being done to the natural environment and to large swathes of the living world that are vital for our health and welfare. The role of the engineer today is therefore every bit as pioneering as it was for our predecessors—but in a very different way.

It is not the purpose of this book to define moral boundaries, or policies set down by governments, but it is appropriate to point out the changing nature of what a 21st-century engineer will be increasingly called upon to do. One of those tasks is to ensure a level of specialized advisory options, consideration of which includes economic and environmental concerns—because these are the drivers of tomorrow, not the ever-gratuitous surge towards increasingly sophisticated tools for conspicuous consumption. Rather the reverse: for engineers must be aware of factors outside the physical laws that have defined their inventions for the entire duration of the scientific process. It is also now necessary for the engineer to be aware of the financial implications of specified projects.

In this more holistic approach to electromechanical engineering, the investment costs balanced by economic factors are increasingly sought by governments and corporations, and the engineer is frequently in a unique position to give qualitative and quantitative advice: for instance, what is the cost–benefit ratio of a switch to battery-powered EVs rather than those powered by fuel cells? Such a question can only be answered with detailed technical knowledge about each step in the conversion process. Those challenges and the recommendations that ensue are sometimes at variance with priorities demanded by the corporate interests of major engineering companies. Nevertheless, holistic solutions to separate problems can apply a systems engineering approach to the answer in a similar way to that which pervaded the fast-track development of major engineering projects during the Cold War, and to which we have referred in several places in this book.

An example exists to demonstrate this. There is a need to replace the fossil fuel-fed ICE in road vehicles and governments are also keen to replace hydrocarbon fuels used for domestic water heating systems. When viewed individually, these challenges attract dedicated evaluation, each separate with solutions tailored to the needs of each sector. The fuel-cell car dispenses with the toxic lithium-ion battery but is far more expensive, both in development and in drawing down the hydrogen reactant for ready availability. But its practicality is unquestioned: a fuel-cell car has a greater range than a conventional gasoline car, takes less time to fill up with fuel (hydrogen rather than gasoline), and has a longer mean time between failures. Moreover, a fuel-cell vehicle maintains a consistent drawdown on power during use, unlike a battery-powered EV, which exhausts its energy source unless a parallel hybrid of the type described in Chapter 12 is adopted.

There is increasing concern among the general engineering fraternity that established principles regarding decisions and recommendations are made increasingly as a result of popular demand or political expediency. Moreover, it frequently appears to those who make difficult decisions at board level that neither the politician nor the public is aware of the balance between investment and profitability. This is the reason why the "hydrogen economy," serving to provide fuel for electric motor vehicles, the transport industry and aviation and which can also be made to provide energy in the home, replacing hydrocarbon energy, is marginalized. For industry moguls, it would cost more to convert all our energy needs to hydrogen in the short term but far less in the long term—and it would be much more environmentally sustainable to do so.

On the one side, captains of industry will always choose the optimum investment for the highest profit, while governments lack the technical knowledge to make choices based on long-term value or environmental sustainability. On the other side, subsidies paid by governments for toxic battery-run cars using cheap labor from poverty-stricken countries prevent the development of a more sustainable hydrogen economy that would replace all hydrocarbon use with one product that, on the proven record of capitalism, will provide a level of demand that would collapse the higher price of transport and domestic heating requirements. Nevertheless, despite these serious challenges, it is the job of the engineer to design a future with responsible analysis based on knowledge and experience of having transformed the world over these past 200 years. It is also his and her job to educate the public in a responsible and rational adaptation of society for the betterment of the human condition everywhere.

GLOSSARY

1. *AC (alternating current)*, an electric current which periodically reverses direction and changes its magnitude continuously with time, the form of electric power delivered to businesses and residences.

2. *Accumulator*, an energy storage device which accepts energy, stores it, and releases it when required. Some accept energy at a high rate over a short time interval and deliver at a low rate over a long time.

3. *Archimedean screw*, a device for lifting water, grains, or powder from a lower-lying deposit to a higher level by rotating a screw-shaped surface.

4. *Artificial Intelligence (AI)*, more commonly called machine intelligence in engineering circles, essentially any device that perceives its environment and takes actions that maximize its chances of satisfactorily achieving its assigned goals.

4. *Bacon cell*, see *Fuel cell*

5. *Biomechanics*, a study of the structure, function, and motion of the mechanical aspects of biological systems at a level from whole organs to organisms using the methods of mechanics.

6. *Blast furnace*, used for smelting, the name of which refers to the combustion air forced into the crucible above atmospheric pressure with fuel ores fed into the top with the air fed in through the bottom to induce chemical reactions.

7. *Boolean logic*, a primary data type within a computer logic which reflects the binary logic of logic gates and transistors in a central processing unit. Can also have its root in Boolean algebra itself at the heart of which is the idea that all values are either true or false.

8. *Bowden cable*, a flexible cable used to transmit mechanical force by the movement of an inner cable through an outer sheath.

9. *Brass*, an alloy of zinc and copper in various proportions for different mechanical and electrical properties and usually used for fashioning tools or small structures. Attractive for its visually esthetic appeal in domestic applications.

10. *Brayton cycle*, a thermodynamic cycle that describes the working of a constant-pressure heat engine.

11. *CAD (Computer Aided Design)*, the use of computers to aid in the creation, modification, or analysis of a design by software used by a designer to create a database often in the form of electronic output.

12. *CAM (Computer Aided Manufacture)*, a computer-assisted manufacturing tool usually generated in CAD and verified through Computer Aided Engineering (CAE).

13. *Camshaft*, a rotating object which carries pointed cams and which is designed to convert rotational motion to a reciprocal action, mainly used in internal combustion engines.

14. *Cast iron*, a group of iron-carbon alloys with a carbon content of more than 2%, possessing a low melting temperature, with quantities of silicon and manganese.

15. *Cerrusite*, a mineral consisting of lead carbonate, an ore, and a secondary mineral of lead formed by carbonated water acting on the mineral galena.

16. *Compression*, the loading or principal effect in squeezing or shortening the component, a reduction in volume and increase in density of a substance under more pressure.

17. *Commutator*, a rotary electrical switch applied to some types of electric motor and electrical generators that periodically reverse the direction of the current between the rotor and the external circuit.

18. *Condenser*, designed to transfer heat from a working fluid to a secondary fluid or to the surrounding air, relying on efficient heat transfer that occurs during phase changes.

19. *Conductive reasoning*, bridging the empirical and scientific methods in which observation and natural deduction result in a conclusion, a process which frequently precedes the application of the scientific process.

20. *Coulomb's law*, or inverse-square law as an experimental law of physics that quantifies the amount of force between charged objects, the force between charged bodies at rest is known as electrostatic force or Coulomb force.

21. *Crankshaft*, a rotating shaft which, in conjunction with connecting rods, converts reciprocating motion of pistons into rotational motion, commonly used in internal combustion engines.

22. *DC (direct current)*, a unidirectional flow of electricity which may be made to flow through a conductor such as a wire but which can also flow through semiconductors, insulators, or through a vacuum as in electron or ion beams.

23. *Diesel engine*, an internal combustion engine in which ignition of the fuel is caused by the elevated temperature of air in a cylinder due to mechanical compression.

24. *Differential*, a gear train with three shafts where the rotational speed of one shaft is the average speed of the others, or a fixed multiple of that average.

25. *Dynamo*, an electrical generator that creates a direct current (DC) using a commutator and was the foundation for successive developments including the electric motor and the alternator.

26. *Electromagnetic spectrum*, covers electromagnetic waves with frequencies ranging from 1 Hz to above 10^{25} Hz corresponding to wavelengths from several thousand km to a fraction of the size of an atomic nucleus.

27. *Empirical Method*, an activity or thought process derived from experience or by experiment without the use of the scientific method.

28. *Fly By Wire (FBW)*, a system that replaces the conventional manual flight controls of an aircraft with an electronic interface where the movement of the controls are converted to electronic signals transmitted by wires.

29. *Force*, defined in mechanical engineering as an action that maintains the motion of a body, one of Isaac Newton's three laws of motion.

30. *Friction*, a force between two surfaces sliding across each other, a resistance to motion when two objects are in contact.

31. *Fuel cell*, any class of a device which converts chemical energy into electricity by electrochemical reactions, resembling a battery but capable of supplying energy over a much longer period of time.

32. *Gear*, in engineering a rotating machine, part of which has cut teeth or cogs which mesh with another toothed part to transmit torque, usually as a mechanism for changing speed, torque, or power transmission rates.

33. ***Horsepower***, a unit of measurement relating to power output defined by the rate at which work is done by engines or motors and based on the estimated work output of a pony.

34. ***Hydraulics***, involves chemistry and physics to manipulate and use the mechanical properties of liquids and is considered a counterpart of pneumatics.

35. ***Hysteresis***, the dependence of the state of a system on its history, for instance where a magnet may have more than one possible magnetic moment depending on how the field changed in the past.

36. ***Induction motor***, usually run on AC single-phase or three-phase power, a few on two-phase. The electric current in the rotor needed to produce torque is obtained via electromagnetic induction from the rotating magnetic field of the stator winding.

37. ***Internal combustion engine (ICE)***, any one of a group of devices in which combusted reactants (oxidizer and fuel) serve as an engine's working fluid, work resulting from the hot gases acting on moving parts.

38. ***Internal energy***, as applied to a thermodynamic system, is a measure of the energy within it, excluding the kinetic energy of the system as a whole.

39. ***Inverter***, a power electronic device of circuit that changes DC current to AC current, the input voltage, output voltage, and frequency as well as overall power handling depending upon the design of the specific device.

40. ***Kinetic energy***, a form of energy that an object or a particle has by reason of its motion, a property applied to a moving object by its motion and its mass.

41. ***Mean Time Between Failure (MTBF)***, measures the average amount of time that equipment is measured to operate between breakdown, failure, or stoppage, and noted in hours of elapsed time. Helps businesses market their products for a reliability standard.

42. ***MGU-H***, Motor Generator Unit-Heat, producing electricity in a motor vehicle from the hot gases of a turbine and stores that in a battery eliminating turbo lag.

43. ***MGU-K***, Motor Generator Unit-Kinetic: converts kinetic energy in a motor vehicle generated under braking into electricity.

44. ***Mortise and Tenon***, the former being a socket or recess, the latter an extended tongue shaped to fit inside the cavity of the mortise to create a strong joint.

45. ***Nanoengineering***, engineering on the nanoscale, deriving its name from the unit of measurement indicating a billionth of a meter with emphasis on the engineering rather than the science of it.

46. ***Nuclear fission***, the process of splitting a large atomic nucleus into smaller nuclei, usually in a nuclear reactor where a neutron is absorbed into a nucleus which causes the nucleus to become a different isotope and unstable.

47. ***Nuclear fusion***, a nuclear reaction where two or more atomic nuclei are combined to form one or more atomic nuclei and subatomic particles, the difference in mass between the reactants and the products releasing or absorbing energy.

48. ***Nuclear propulsion***, a hypothetical means of producing motive power from heating a fuel by nuclear fission reactor to produce a reactive thrust in a rocket motor.

49. ***Nuclear reactor***, a device that can initiate and control a self-sustaining series of nuclear fissions for the purpose of driving a turbine to produce electricity.

50. ***Otto cycle***, a description of what happens when a mass of gas is subjected to changes of pressure, temperature, volume, added heat, and the removal of heat, essentially defining the four-stroke cycle of an internal combustion engine.

51. ***Pneumatics***, a branch of engineering utilizing the use of pressurized gas or air to power systems through the use of centrally located compressors for motors, cylinders and other pneumatic devices.

52. ***Powertrain Control Module (PCM)***, a control unit on motor vehicles consisting of an engine control unit (ECU) and the transmission control unit (TCU) and frequently involving several small computers integral to the vehicle.

53. ***Programmable logic controller (PLC)***, an industrial digital computer ruggedized for the control of manufacturing processes.

54. ***Rotary engine***, an internal combustion engine in which the cylinders are arranged radially and rotate around the central crankshaft which remains stationary with the entire cylinder block rotating around it.

55. ***Scientific Method***, deduction obtained through calculation, experiment, or research, sometimes also through test using real-world or simulated examples.

56. ***Shear force***, an internal force in any material usually caused by an external force acting perpendicular to the material or a force which has a component acting tangent to the material.

57. ***Sigma***, a scale of how well a critical characteristic performs compared to its requirement, the higher the numerical score the more capable the characteristic.

58 ***Smelting***, a process involving a blast furnace to produce pig iron which is then converted into steel, with carbon acting as a reactant to strip oxygen from the ore.

59. ***Tension***, a transmitted force defined by an action-reaction influence and sometimes referred to as the reciprocal of compression. It relates to the forces exerted by the object to which opposing ends of a string or structures are attached.

60. ***Torsion***, the twisting of an object resulting from an applied torque exerted at either end imparting a twisting motion with shear stress by torque on a component.

62. ***Torque***, a moment of force defined by the tendency of a force to rotate a body to which it is applied and specified by the axis of rotation, the component force vector lying in the plane perpendicular to that axis.

63. ***Turbine***, a rotary mechanical device that extracts energy from a fluid and converts it into useful work.

64. ***Turboprop engine***, a turbine engine that drives an aircraft propeller for forward motion. Air is drawn in and compressed before adding fuel for combustion, expanding through a turbine.

65. ***Vector***, in physics and mathematics, any quantity with both a magnitude and a direction but when applied where there is velocity, it describes other quantities such as acceleration and momentum, which are also vectors.

66. ***Velocity***, the object's speed and direction of motion, a basic concept in kinematics describing the motion of bodies, a physical vector quantity defined by both magnitude (speed relative to a fixed base) and direction.

67. ***Worm gear***, the geared teeth on a rotating wheel aligned with a helical screw transmitting rotational motion to linear motion. Gear reductions of up to 300:1 can be effected but while the worm can turn the gear, the gear cannot turn the worm.

SUGGESTED READING

CHAPTER 1: THE DAWN OF IDEAS—THE EMPIRICAL AGE

Lyon Sprague, De Camp (1963). *The Ancient Engineers.* Doubleday. A generalized examination of the methods and techniques used in the ancient world.

Oleson, John Peter (1984). *Greek and Roman Mechanical Water Lifting Devices. The History of Technology.* Dordrecht: D. Reidel. Examples of early engineering applications with descriptions of different solutions to a common problem.

Oleson, John Peter (2009). *Oxford Handbook of Engineering and Technology in the Classical World.* Describes empirical thinking in the evolution of tools and machines during the period dominated by Greece, Rome, and the Islamic world.

White, Jr., Lynn. (1962) *Medieval Technology and Social Change.* Oxford at the Clarendon Press. A good overview of the impact of empirical engineering on the structure of society and societal behavior.

CHAPTER 2: UNDERSTANDING FORCES—THE SCIENTIFIC AGE

Corben, H. C.; Stehle, Philip (1994) *Classical Mechanics.* Dover Publications. Provides an underpinning set of rationales for the definition and applications of terms, phrases, and calculations as they apply to mechanical engineering.

Kleppner, Daniel; Kolenkow, Robert J. (2010). *An Introduction to Mechanics.* Cambridge University Press. A guide to fundamentals in forces and structures.

Serway, Raymond A. (2003). *Physics for Scientists and Engineers.* Philadelphia: Saunders College Publishing. The underlying principles and equations in physics.

CHAPTER 3: MANIPULATION OF FORCES

Bejan, Adrian (2016). *Advanced Engineering Thermodynamics.* Wiley. A good follow-up to preliminary and introductory texts but one which is better appreciated with a grounding in thermodynamic theory and mathematics.

Brown, Richard (2002). *Society and Economy in Modern Britain 1700–1859.* Taylor & Francis. A social and societal analysis of the impact of the industrial age on life in general and the consequences of engineering development.

M. Scott Shell (2015). *Thermodynamics and Statistical Mechanics: An Integrated Approach.* Cambridge University Press. An advanced text which provides a step up to advanced engineering theory approached by way of fundamental principles.

CHAPTER 4: MECHANISMS

J. J. Uicker, G. R. Pennock, and J. E. Shigley, *Theory of Machines and Mechanisms*, Oxford University Press. A good, sound background and reference work defining the operating principles and the application of theory to practice.

Lung-Wen Tsai (2001), *Mechanism design, enumeration of kinematic structures according to function.* CRC. A good description of the application of mechanical principles to working machines and the way structures can be made to converge at an interface.

CHAPTER 5: A CHOICE OF MATERIALS

Ashby, Michael; Hugh Shercliff; David Cebon (2007). *Materials: engineering, science, processing and design.* Butterworth-Heinemann. A background to how materials relate to their engineering application through their structure.

Askeland, Donald R.; Pradeep P. Phulé (2005). *The Science & Engineering of Materials.* Thomson-Engineering. How materials relate to each other, applied through technology, and how they can be engineered for unique applications.

Walker, P., ed. (1993). *Chambers Dictionary of Materials Science and Technology.* Chambers Publishing. A general reference work which also contains some useful parameters and descriptions of material applications.

CHAPTER 6: KEEPING THE MACHINES MOVING

Heeresh Mistry (2013) *Fundamentals of Pneumatic Engineering* (2013). Create Space e-Publication. A basic text book of pneumatic engineering with some examples of applications and uses.

Ohanian, Hans (1994). *Principles of Physics.* W.W. Norton and Co. Contains some useful material associated with the Carnot theorem.

Moran, Michael J., and Howard N. Shapiro (2008). *Fundamentals of Engineering Thermodynamics.* John Wiley. Addresses the fundamental principles underpinning the design and development of the internal combustion engine.

Nunney, Malcolm J. (2007). *Light and Heavy Vehicle Technology.* Elsevier Butterworth-Heinemann. A good grounding in the engineering and technology applied to a wide range of land vehicles.

Terry S. Reynolds (2002). *Stronger Than a Hundred Men: A History of the Vertical Water Wheel.* JHU Press. The convergence of empirical and scientific thinking in tackling universal challenges using technology and engineering.

CHAPTER 7: ENGINEERING THE SYSTEMS

Blockley, D. Godfrey, P. (2017). *Doing it Differently: Systems for Rethinking Infrastructure.* Ice Publishing. Options in strategy, systems engineering, and planning for complex engineering solutions.

Daniel Kandray (2010). *Programmable Automation Technologies*, Industrial Press. A good guide to the range of automation technologies and the manner in which electronic interface can leverage the value of mechanical systems.

Duffy, James E. (2015). *Modern Automotive Technology.* Goodheart-Wilcox. An excellent primer for all aspects of motor vehicle engineering and highly recommended as a career-guide for the motor vehicle maintenance industry.

Klastorin, Ted (2003). *Project Management: Tools and Trade-offs.* Wiley. How to approach the management of engineering, technology, and systems from a management perspective with some of the pitfalls and benefits.

CHAPTER 8: WHAT IS RELIABILITY?

Federal Aviation Administration. (2013). *System Safety Handbook*. FAA. An ideal briefing and safety standards manual for aeronautical engineers, quality control inspectors/managers, and for understanding the rationales behind legislation and standards across all applications of engineering.

Juran, Joseph and Gryna, Frank. (1988). *Quality Control Handbook.* McGraw-Hill, New York. Although somewhat old, this book carries all the fundamental criteria for quality control, reliability, and the necessary tools for constructing engineering projects to maximize safety and reliability.

Theodore Hailperin. (1986). *Boole's Logic and Probability: a critical exposition from the standpoint of contemporary algebra, logic, and probability theory*. Elsevier. A detailed explanation and analysis of Boolean applications in systems engineering, useful for application to fault-tree calculation.

CHAPTER 9: NUCLEAR ENGINEERING

Gowing, Margaret. (1964). *Britain and Atomic Energy, 1939–1945*. HMSO. An engrossing development history of how the UK came to be a world leader in nuclear engineering and did much of the pioneering work on nuclear energy.

Shultis, J.K. & Faw, R.E. (2002). *Fundamentals of Nuclear Science and Engineering*. CRC Press. A basic textbook of nuclear physics with a view to its application in nuclear engineering, largely for power production.

CHAPTER 10: ELECTRICAL AND ELECTRONIC ENGINEERING

Avallone, Eugene A; et al., eds. (2007). *Marks' Standard Handbook for Mechanical Engineers*. McGraw-Hill. Introductory reference work presented in a readable style but useful for its notations and for displaying the several areas of speciality for mechanical engineers.

Bakshi, U. A.; Godse, A. P. (2009) *Basic Electronic Engineering*. Technical Publications. An essential introductory guide to the basics of the subject with good explanatory text and recommended as a starting point for this area of mechanical engineering.

Horowitz, Paul and Hill, Winfield. (2015). *The Art of Electronics*. Cambridge Univ. Press. Provides good coverage of the subject at an understandable level for the layman and an appropriate introduction to the subject with reference to different aspects of this field.

Bhargava, N. N. and Kulshrishtha, D. C. (1984). *Basic Electronics and Linear Circuits*. Tata McGraw-Hill Education. A good background to the subject, with an explanation of the different kind of circuits. Good on the theory of electronics.

CHAPTER 11: POWER PRODUCTION AND GENERATION

Brodie, David; Brown, Wendy; Heslop, Nigel; Ireson, Gren; Williams, Peter (1998). *Physics*. Addison Wesley Longman Limited. The physical forces involved in motive power and the application of physical law to machines.

Tagare, Digambar M. (2011). **Electric Power Generation: the changing dimensions.** Wiley. An analytical overview of the current electricity generating processes together with coverage of hydroelectric, thermal, and nuclear systems. Also covers renewables and pv cells.

Jain, Mahesh C. (2009). *Textbook of Engineering Physics*. New Delhi: PHI Learning Pvt. Ltd. Takes the basic principles of physical law and applies them to the science and technology involved in their application to engineering.

Lockwood, Thomas D. (1883). *Electricity, Magnetism, and Electric Telegraphy*. D. Van Nostrand. A good background to the travails that usually attach to the development of a new technology.

Vielstich, W.; et al., eds. (2009). *Handbook of Fuel Cells: advances in electrocatalysis, materials, diagnostics and durability*. Hoboken: John Wiley and Sons. Defines the working principles and the mechanical advantage of the fuel cell and discusses technical issues of their design, development, and application.

CHAPTER 12: ELECTROMECHANICS

Lindsay, J. F.; Rashid, M. H. (1986). *Electromechanics and Electrical Machinery*. Prentice-Hall. Integrates the principles of eletromechanics and electrical engineering as applied to the technology and the machines it powers.

Miu, Denny K. (2011). *Mechatronics: Electromechanics and Controlmechanics*. Springer-London. Covers several specialized aspects of electromechanics and their control systems but a good guide to some of the more exotic applications.

Skilling, Hugh Hildreth (1960). *A First Course in Electromechanics*. Wiley, 1960. Sound principles of the entire field of electromechanics with more on the physics and mathematics without recourse to specific projects.

CHAPTER 13: COMPUTING MACHINES AND ARTIFICIAL INTELLIGENCE

Fung, Y.-C. (1993). *Biomechanics: Mechanical Properties of Living Tissues*. New York: Springer-Verlag. A technical explanation for the field with relevant applications, pointing the reader to solutions for mechanical devices to limb amputees, etc.

Ifrah, Georges (2001). *The Universal History of Computing: from the abacus to the quantum computer.* New York: John Wiley & Sons. An extensive background history of computing linking theory with practical applications and a peer into the future with quantum computers.

Johnston, John (2008) *The Allure of Machinic Life: Cybernetics, Artificial Life, and the New AI*. MIT Press. A pointer to possibilities and to future potential for AI and cybernetics, with a prediction for developmental paths and ways in which AI could be interacted with machines for the benefit of humans.

Stokes, Jon (2007). *Inside the Machine: An Illustrated Introduction to Microprocessors and Computer Architecture*. San Francisco: No Starch Press. An introductory and easy-to-read explanation of the mechanical integration of computers and microprocessors.

Sun, R. & Bookman, L. (eds.) (1994). *Computational Architectures: Integrating Neural and Symbolic Processes*. Kluwer Academic Publishers, Needham, MA. Although somewhat dated by publication, this is a good stimulus for creative thinking about the future and the way SI, computational processes, and machine learning can proceed through a range of different architectures.

INDEX